The Communicative Engine

The Communicative Engineer: How to Ask, Listen, Write, Speak, and Use Visuals

Stuart G. Walesh, PhD, PE

WILEY

Published by John Wiley & Sons, Inc., Hoboken, New Jersey.
Published simultaneously in Canada.

For general information on our other products and services or for technical support, please contact our Customer Care Department within the United States at (800) 762-2974, outside the United States at (317) 572-3993 or fax (317) 572-4002.

Wiley also publishes its books in a variety of electronic formats. Some content that appears in print may not be available in electronic formats. For more information about Wiley products, visit our web site at www.wiley.com.

Library of Congress Cataloging-in-Publication Data applied for:

Paperback: 9781394202591

Cover Design: Wiley
Cover Image: © Javier Zayas Photography/Getty Images

SKY10067314_021724

To Jerrie, my wife,
always graciously and competently helpful

Contents

Preface

PURPOSE

Practicing engineers spend about half of their work time communicating, that is, asking, listening, writing, speaking, and using visuals to convey ideas, information, and feelings from one person to one or more others. They ask colleagues for technical help, answer probing questions from demanding clients, write major reports that urge the implementation of costly recommendations, speak about controversial environmental issues at national conferences, and prepare images to illustrate complex processes. There is some truth in the stereotypical image of engineers; that is, too many engineers communicate poorly. That deficiency can be corrected with this book's multi-modal approach—asking, listening, writing, speaking, and using visuals.

Furthermore, while communication knowledge, skills, and attitudes (KSA) have always been an essential part of engineering practice, communication's importance has increased in recent decades. Various forces drive the need for enhanced communication within all engineering disciplines. Some examples include increased public access, via social media and the press, to information about environmental and infrastructure issues; a more informed, concerned, and engaged public; explosion of regulations; societal divisiveness; fear of crime and terrorism; and global uncertainties. This increased complexity trend will continue, and all aspects of it have engineering components and require communicative engineers.

Accordingly, for the benefit of society, more of today's and tomorrow's engineers should complement their technical competency with communication competency in order to function and sometimes lead in an increasingly complex socio-economic-political environment. That is the motivation for writing this book. More specifically, its purpose is to:

- **Demonstrate how effective communication results in successful engineering projects and other engineering endeavors:** Engineering students and practitioners should understand the critical role of communication in project success. Miscommunication frequently produces failures resulting in fatalities, injuries, and property and environmental destruction.
- **Describe effective communication as drawing on six communication modes**: Asking, listening, writing, speaking, visuals, and mathematics—to convey ideas, information, and feelings.

- **Show how to apply the first five modes, using hypothetical and actual engineering situations:** The intended result: communicative engineers; a safer society; and better-served clients, customers, and stakeholders.

The book's premise is that the most highly successful engineers across all disciplines, as measured by value added, societal impact, and personal gain, are those with technical and non-technical KSA sets. Communication is the prime non-technical component.

AUDIENCES: STUDENTS AND PRACTITIONERS

I wrote *The Communicative Engineer: How to Ask, Listen, Write, Speak, and Use Visuals* with the assumption that readers are primarily undergraduate or graduate students in engineering. The secondary audience is engineering practitioners and students or practitioners in similar technical and scientific communities. The book assumes readers are receptive to a proactively build communication competency to complement their evolving technical/scientific competencies.

Instructors might select this book as the text for a comprehensive communication course. Another or supplemental approach would be for faculty to expect students to use the book as a resource for many of their undergraduate and graduate courses, as they proceed through the curriculum from their first year of engineering study on to their ultimate degree goal. This text could be the resource that supports a communication-across-the-curriculum program.

Students, beginning with their first year of college, will find much of the material in this book immediately useful. It will help them complete writing assignments in their engineering and non-engineering courses; prepare for presentations to classmates and faculty; interact with team members; be more productive during internships and summer jobs; interview with prospective employers; and clarify their thinking about important topics and issues.

I hope students will find the book to be so helpful that they will continue to use it as engineering practitioners. Given the book's foundation of communication fundamentals, I believe that it will also be applicable outside of engineering—especially to student and practitioner members of the larger scientific-technical-business community.

Speaking of practitioners, engineers in practice can use *The Communicative Engineer* in a just-in-time manner. For example, when asked to lead the writing of a report for your project team, go immediately to Chapter 3, "Writing," or, if offered a speaking opportunity, explore Chapter 4, "Speaking," for guidance.

Instead of working methodically through the book in a chapter-by-chapter manner, students could use the book while in college in a "pop in and pop out" manner. I designed it so students and practitioners can find what they need when they need it. *The Communicative Engineer* offers a detailed index along with a table of contents that includes many headings and subheadings. The text often refers to other relevant sections, which directs readers to additional useful material.

This book will prove useful in supporting the internal education and training programs of business, public, and academic sector employers of engineers in all disciplines. Engineering societies, such as the five founder societies, which are the American Institute of Chemical Engineers (AIChE), the American Institute of Mining, Metallurgical, and Petroleum Engineers (AIME), the American Society of Civil Engineers (ASCE), the American Society of Mechanical Engineers (ASME), and the Institute of Electrical and Electronic Engineers (IEEE), can confidently use this book as the foundation of their communication programs.

The Communicative Engineer will also strengthen the continuing education programs of associations that include engineers and non-engineers. Some examples are the American Public Works Association (APWA), the Association for Computing Machinery (ACM), and the Energy Management Association (EMA). The book's fundamentals, enhanced with tailored versions of its examples and exercises, will support communication webinars, seminars, and workshops attended by practicing engineers and other technical and non-technical personnel.

ORGANIZATION AND CONTENT

Chapter 1 defines communication and describes its benefits for individual engineers, their employers, and the individuals and entities they serve. It also illustrates the costs, monetary and otherwise, of poor or failed communication. The chapter introduces the five modes of communication—asking, listening, writing, speaking, and use of visuals—and establishes principles applicable to essentially all modes. The text argues that engineers are poised to be good to great communicators. Finally, as a means of encouraging and inspiring readers, Chapter 1 introduces some exemplary engineer communicators and shares some of their statements.

Chapter 2 addresses the related asking and listening modes, which are the basis for effective interpersonal communication, and offers practical tips. With those two basic modes as a starting point, Chapters 3, 4, and 5 address, respectively, the more complex writing, speaking, and visual communication modes. After describing the fundamentals of a mode, its chapter shows, in pragmatic fashion, how to apply those fundamentals using hypothetical and actual engineering situations. *The Communicative Engineer* frequently reminds readers that, ultimately, they are responsible for developing their communication KSA.

Each chapter begins with a list of learning objectives—that is, what I hope you will learn and be able to do after working through the chapter. Chapters include many true personal and other stories to illustrate the text's ideas and content.

Chapters conclude with a list of key points followed by cited references and exercises. The numbering system used with chapter headings and subheadings (e.g., 1.5 and 1.5.2) enables helpful back-and-forth references within a chapter or to any other part of the book.

The Communicative Engineer occasionally offers communication guidance and advice specifically for practicing engineers. These practitioner suggestions are also of value to engineering students because they see how the fundamentals presented in the book are ultimately applicable to practice.

Exercises, which appear at the end of all chapters, provide opportunities to use communication fundamentals and techniques presented in the chapters. Many exercises are well suited for modest to major team projects. Teamwork, especially when the teams are composed of cognitively diverse individuals, stimulates using and becoming more proficient with various forms of communication. Therefore, instructors could assign some exercises as team projects. In that way, students will learn more about the subject matter and become even better communicators and team members.

The book's seven appendices provide supplemental material. Topics: abbreviations, introductions to communicative engineers, suggested questions, style guide ideas, punctuation guidelines, examples of specific types of communication, and illustrations of speaker liabilities.

ACKNOWLEDGMENTS

Many accomplished and varied individuals kindly assisted me in meeting the book research and writing challenges by questioning some of my assertions, suggesting and/or providing resources, outlining additional key ideas, offering book organization and format ideas, clarifying and tightening text, and answering questions. Collectively, they reflect the views of engineers and others in the academic, business, and government sectors.

I am indebted to the following for their assistance: Greg Adamson, Tomasz Arciszewki, Rachelle Leigh Beckner, Danielle Boykin, Allen Estes, Marco Fellin, Larry Galler, Robert Green, Neil Grigg, Chip Kilduff, Tom Lenox, Chris Kaufman, Kenton Machina, Maggie Miles, Chad Morrison, Henry Petroski, Steve Polcyn, Simine Short, David Soukup, Kassim Tarhini, and Ted Weidner. Each one helped me in one or more ways write with what I hope is credibility and value to students and practitioners. However, I am totally responsible for the manner in which I have used their contributions.

I acknowledge and appreciate what I have learned from and with students, practitioners, and others who took my university courses or participated in my seminars, webinars, and workshops. Anonymous professionals who reviewed my proposal for this book and reviewed the draft manuscript added value, for which I am indebted. The many and varied sources cited in this book illustrate my debt to many individuals and organizations. I drew ideas, information, and reference materials from a wide range of sources.

The contributions of members of the Wiley team are appreciated, especially Kalli Schultea, Editor, Civil Engineering and Construction; Vishal Paduchuru,

Managing Editor; Isabella Proietti, Editorial Assistant; and Kavya Ramu, Content Refinement Specialist.

Finally, Jerrie, my wife, carefully proofed punctuation, spelling, and grammar; critiqued content; told me when I was preaching or beating around the bush—and, as always, provided total support.

Stuart G. Walesh
Valparaiso, IN
12 January 2024

About the Author

Stuart G. Walesh, PhD, PE, Dist.M.ASCE, F.NSPE, practicing as an independent consultant-teacher-author, provides management, engineering, and education/training services for business, government, academic, and volunteer sector organizations. He earned a BS in civil engineering from Valparaiso University, a master's degree in engineering from Johns Hopkins University, and a PhD in engineering from the University of Wisconsin-Madison. He is a licensed professional engineer.

Stu has over five decades of engineering, education, and management experience in government, academic, and business sectors. He served as a project manager, department head, discipline manager, author, marketer, sole proprietor, instructor through professor, and dean of an engineering college. As a member of various organizations, Stu coached junior professionals in areas such as communication, team essentials, project planning and management, and affecting change.

Water resources engineering is Stu's technical specialty. Over the years, he led or participated in watershed planning, computer modeling, flood control, stormwater and floodplain management, groundwater, dam, and lake projects. His engineering experience includes project management, research and development, design, stakeholder participation, litigation consulting, and expert witness services. Areas in which he provides management and leadership assistance include education and training, mentoring, research, writing and editing, speaking, marketing, meeting planning and facilitation, project planning, and team essentials.

In addition to this book, Stu authored:

- *Urban Surface Water Management* (Wiley 1989)
- *Flying Solo: How to Start an Individual Practitioner Consulting Business* (Hannah Publishing 2000)

- *Managing and Leading: 52 Lessons Learned for Engineers* (ASCE Press 2004)
- *Managing and Leading: 44 Lessons Learned for Pharmacists* (co-authored with Paul Bush, American Society of Health-System Pharmacists 2008)
- *Engineering Your Future: The Professional Practice of Engineering* (Wiley 2012; the first and second editions were published in 1995 and 2000)
- *Creativity and Innovation for Engineers* (Pearson 2017)
- *Engineering's Public-Protection Predicament* (Hannah Publishing 2021)

He also authored or co-authored hundreds of publications and presentations about engineering, education, and management, and facilitated or led workshops, seminars, webinars, and meetings throughout the United States and internationally.

Stu served on or led various professional societies and community groups. Over the past two decades, he has been active in the effort to reform the education and early experience of engineers. During the past decade, Stu has studied, written, and spoken about how to use recently discovered basic brain knowledge to help individuals and their teams work smarter—that is, be more effective, efficient, and creative/innovative. More recently, he challenged the American engineering community to remove the dichotomy between engineering's ethics codes, which claim public protection is paramount, and the widespread licensure-exemption laws in which massive amounts of engineering are conducted without the guidance of state-licensed engineers, thus placing the public at unnecessary risk.

His professional work and service to society have been recognized by the American Society for Engineering Education, the Consulting Engineers of Indiana, the American Society of Civil Engineers, the Indiana Society of Professional Engineers, the National Society of Professional Engineers, the University of Wisconsin, and Valparaiso University.

For additional information, go to www.HelpingYouEngineerYourFuture.com or contact him at stu-walesh@comcast.net.

CHAPTER 1

INTRODUCTION

Communication is not what is intended,
but what is received by others.

—Mel Hensey, consultant

After studying this chapter, you will be able to:

- State this text's three-part purpose
- Explain engineering's ideology and the related importance of effective communication
- Define communication
- Describe benefits effective communication provides to individual engineers, engineering organizations, and the public
- Provide examples of the monetary and other costs of poor communication
- Discuss principles of effective communication
- Identify some notable engineer communicators

1.1 MOTIVATION FOR AND PURPOSE OF THIS BOOK

1.1.1 Hyatt Regency Hotel Tragedy

On July 17, 1981, 114 people died and over 200 were injured when two walkways suspended in the atrium of the Kansas City, MO, Hyatt Regency Hotel collapsed. Plans called for the walkways supported by beams, each connected at one end to a rod attached to the ceiling.

During walkway construction, a member of the construction team, on determining that the original beam-rod connection design was impracticable to construct,

The Communicative Engineer: How to Ask, Listen, Write, Speak, and Use Visuals, First Edition.
Stuart G. Walesh. Published 2024 by John Wiley & Sons, Inc.

called a member of the engineering design team and suggested a modification. The engineer, replying over the phone, indicated that the modification seemed acceptable from a structural perspective and asked for a formal written change order. The contractor either did not submit the change order or did submit it but the engineer did not effectively evaluate it. The net effect of the flawed communication was a doubling of the loads on all of the upper walkway rod-beam connections, which caused the disastrous collapse [1, 2].

1.1.2 Boeing 737 MAX 8 Disasters

In October 2018 and March 2019, 346 people were killed when two Boeing 737 MAX 8 aircraft crashed in a similar manner, nose down at over 500 miles per hour. Thus began years of agony for survivors, most of whom had to relive the tragedy as they sought justice and compensation.

During the 737 MAX 8 design and testing process, Boeing engineers determined that the aircraft nose tended to pitch up. Therefore, they added software that would automatically push the nose down when needed. However, Boeing communicated this change neither to the Federal Aviation Administration (FAA) nor to pilots. Boeing originally described the change in a draft manual but later removed it to avoid the cost of FAA-required pilot education and training. Accordingly, because of that intentional miscommunication, some pilots did not know what to do when the new system pushed the nose down. They had 10 seconds to react, or the new software would push the nose further down, which it disastrously did in both crashes [3]. (Note: See Appendix A for a list of abbreviations used in this book.)

1.1.3 Undocumented Meeting Conflict

Consider one more communication failure with results not as tragic as the preceding two but still costly in monetary and other ways. An engineering firm designed a flood control project for a community. Soon after the completion of construction, a large rainfall caused flood damage. The community sued the engineering firm claiming negligence.

The law firm that represented the engineering firm in the lawsuit retained me to determine what happened. I eventually discovered the date and subject matter of a meeting at which community and engineering firm representatives discussed design criteria. After that meeting, the engineering firm proceeded with the design. Unfortunately, the engineering firm representatives failed to document the options considered and decisions made at the meeting. Those personnel claimed that, based on decisions made during the undocumented meeting, their engineers designed the facilities for a "moderate" storm, not a "big" storm like the one that occurred. Community representatives contended that, based on the undocumented meeting, the firm should have designed for a "big" storm.

The two parties eventually settled the case at great monetary cost to the engineering firm and a damaged relationship with the community, all for lack of documentation of one meeting.

1.1.4 Book's Purpose

Why cite engineering failures, including two with massive injuries and deaths, before stating the book's purpose? Because, based on the ethics codes of essentially all US-based engineering societies, the ideology of engineering is meeting society's major infrastructure and environmental needs while holding public protection paramount. For example, the ethics code of the National Society of Professional Engineers (NSPE) states, "Engineers, in the fulfillment of their professional duties, shall . . . hold paramount the safety, health, and welfare of the public" [4].

Briefly stated, this book's purpose is to help you, and those you work with or for, communicate effectively so that you are more likely to protect the public and less likely to cause or be part of failures like the preceding. More specifically, the purpose of this book is to:

- **Demonstrate how effective communication results in successful engineering projects and other engineering endeavors.** Engineering students and practitioners should understand the critical role of communication in project success. Miscommunication frequently produces failures resulting in fatalities, injuries, and property and environmental destruction.
- **Describe effective communication as drawing on six communication modes**—asking, listening, writing, speaking, visuals, and mathematics—to convey ideas, information, and feelings.
- **Show how to apply the first five modes, using hypothetical and actual engineering situations.** The intended result: communicative engineers; a safer society; and better-served clients, customers, and stakeholders.

Ideally, the medical profession provides medical care without doing harm, and the legal profession seeks justice within the law, while engineering as a profession strives to meet society's physical and environmental needs while keeping public protection paramount. Medicine's, law's, and engineering's principal reasons for being—their ideologies—are, respectively meeting society's health, justice, and physical and environmental needs with profession-specific other requirements [5]. Effective communication will help engineers fulfill their public-protection-is-paramount promise.

1.1.5 Where There's Smoke There's Fire

Let's explore why the book's purpose suggests that engineers could communicate more effectively. Years ago, I attended an annual conference of the American Society for Engineering Education (ASEE). The keynote speaker told the largely faculty audience that engineers are poor communicators. Is that a fact or a stereotype? The 2004 book *Communication Patterns of Engineers* [6] summarizes literature from the previous four decades about the topic of how well engineers communicate. Well into the book, the authors state, "Many agree that many, if not most, engineers have trouble writing and speaking clearly." Again, stating a fact or stereotyping?

In an article titled "Resilient Engineering Identity," engineering professor Monique Ross states, "In addition to being male, engineer stereotypes include social ineptitude, tinkering, lack of creativity, love of math, poor communication skills, and limited/myopic interests." Note the mention of communication. She says this image "perpetuates unequal patterns of participation in engineering and computing" and argues for discarding it to reflect "this truly admirable, exciting, and engaging field." Reducing stereotypes, including engineers are poor communicators, would make engineering more attractive to a broader cross section of society [7].

Because the preceding communication and other stereotypes have persisted for decades, I conclude, in the spirit of where there's smoke there's fire, that many engineers could be better communicators. That is why I researched and wrote this book.

1.2 COMMUNICATION

1.2.1 Communication Defined and Its Importance in Engineering

Given that this is a communication book, let's define that word for frequent use throughout the text. I reviewed communication definitions provided by the Merriam-Webster, Oxford, and Britannica dictionaries. The essentials of those definitions, combined with my interest in stressing a suite of communication modes, led to the following definition: Communication is the act or process of effectively conveying, from one person to one or more others, ideas, information, and feelings using asking, listening, writing, speaking, visuals, and mathematics.

How much of a practicing engineer's time is spent communicating? The previously mentioned book presents the results of studying documents from the previous four decades dealing with how engineers communicate. The authors show that engineers spend at least half their time communicating" [6].

Researchers at North Carolina State University used a structured survey and informal interviews to learn how working professionals in various professions, including engineering, viewed communication in the workplace. Engineers reported spending 35% of their workweek writing [8]. They were not asked about time invested in other communication modes. However, one can reasonably assume that adding time spent asking, listening, speaking, and using visuals and mathematics would result in engineers communicating roughly half the time.

For more insight into the importance of communication in engineering practice, listen to engineering professor and author Henry Petroski: "Some of the most accomplished engineers of all time have paid as much attention to their words as to their numbers, to their sentences as to their equations, and to their reports as to their designs" (H. Petroski, personal communication, 6 March 2023).

If any of us—engineering students or practicing engineers—want to improve our effectiveness, we should first examine how we spend large amounts of our time. Clearly, communication is a top candidate for improvement.

Having made arguments for the importance of communication, I suggest we—engineering faculty and students—recognize that some students will not initially accept that proposition. Informed by experience, I believe that a few aspiring engineers decide to study engineering because they think the field is essentially all understanding and using various technologies. Therefore, they can avoid the burden of learning how to write and speak, which is obviously required in many other occupations.

Try to change their view by telling more true stories about the various costs of poor communication, like those in Sections 1.1.1–1.1.3. Search for additional studies that reveal how practicing engineers use their time. Urge students to seek part-time employment or intern with an engineering organization and observe engineers.

1.2.2 Origin of the Five Modes

Seeing the six modes in the definition and having already noted the use of five of them earlier in this chapter (Section 1.1.4), you may wonder about their origin as used in this book. Let's go back many decades to a day during the fall semester of my first year in engineering college. Others and I were diligently working on an exercise in an engineering drafting class. The engineering dean abruptly entered our graphics class, apologized for the interruption, and said something like this: "While you are here at the university, develop your communication abilities. Learn how to write and speak and how to use mathematics and graphics." Then he immediately left.

I do not remember his exact words. However, as if it were yesterday, I vividly remember these three aspects of the dean's short visit. He, as leader of the engineering college:

- Believed that communication ability was vital to the practice of engineering
- Viewed effective communicators as being able to draw on multiple modes
- Expected engineering students to study communication as part of their university education

That brief experience prompted me to work on my communication knowledge, skills, and attitudes (KSA) in the dean's four areas, and two I added—asking and listening.

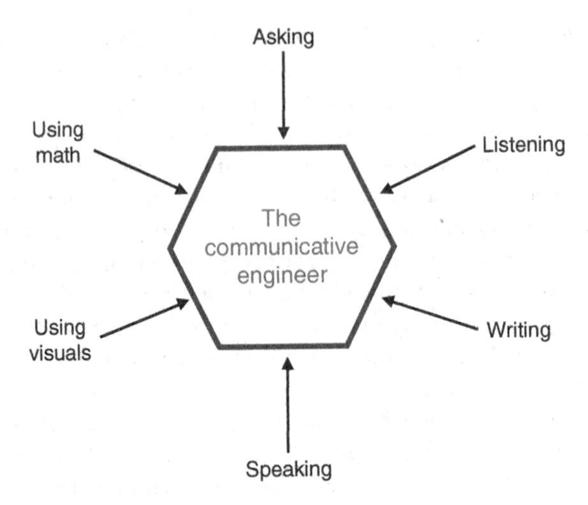

Figure 1.1 The technically competent communicative engineer meets challenges by drawing on various combinations of six communication tools.

Figure 1.1 illustrates my view of the communicative engineer—student or practitioner. This technically competent individual can draw on and effectively use up to six communication modes and produce a safer society, more effective organizations, and better-served clients, customers, and stakeholders.

This book does not discuss mathematics, which the dean mentioned, because engineering curricula address it adequately in mathematics courses and in many courses that apply mathematics. Engineering students know how to use statistics to describe the meaning of data. They can apply probability to provide some insight into the future. Aspiring engineers create digital models of chemical processes.

My commitment to improving my communication KSA has taken me many places, including productive and rewarding positions in academia, business, and government, and travel to countries around the globe. Whatever success and significance I've achieved connects, in part, to continuously improving my communication KSA, and I am still a work in progress. I wish a similar career and life experience for you—that is why I wrote this book and, in it, use the multimodal approach.

1.2.3 KSA Explained

Having introduced KSA, let's illustrate its applicability to the process of improving communication. The NSPE defines knowledge, in KSA, as "comprehending theories, principles, and fundamentals" [9]. Examples of communication knowledge, the understanding of which will enhance your communication effectiveness, are know your audience, whether an individual or group; state your purpose; and accommodate the audience's preferred ways of understanding.

NSPE defines skills as "abilities to perform tasks and apply knowledge." Skill examples related to the preceding knowledge examples are how to profile an audience, how to state your purpose, and how to speak to an audience that includes individuals with varied preferred ways of understanding.

Attitudes, according to NSPE, are "the ways in which one thinks and feels in response to a fact or situation." An example of a constructive communication attitude, based on my many experiences working with editors, would be to receive a heavily edited report back from a professor or supervisor and see it as an opportunity to learn about writing and to improve the document—as opposed to being discouraged or annoyed.

In essence, knowledge means your understanding of powerful fundamentals that change little with time. Skills, which change with time because of technological advances and other forces, are practical things you are able to do with that knowledge. Attitudes describe how you respond to various positive and negative situations.

Looking ahead, the next four chapters provide broad and deep discussions of, respectively, asking and listening presented together in one chapter and writing, speaking, and using visuals each presented in separate chapters.

1.3 BENEFITS OF EFFECTIVE COMMUNICATION

Consider some benefits to you, your employer, and those you serve when you and others strive to be effective communicators.

1.3.1 Enhanced Public Protection

The most important benefit is public protection. The vast majority of engineering projects are successful, sometimes including giving prime attention to public health, safety, and welfare, partly because of effective communication within engineering teams and between teams and those they serve. As suggested by the two tragedies featured at the beginning of this chapter, miscommunication causes some engineering disasters.

Another cause of unnecessary tragedies in the United States is exemptions to state engineering licensure laws. The District of Columbia and all states, except Arkansas and Oklahoma, allow conduct of some risky engineering projects without the guidance of competent and accountable licensed engineers (Professional Engineers, PEs), whose primary responsibility is public protection. NSPE and some individual engineers work to reduce or eliminate exemptions to licensure laws [10].

1.3.2 Improved Organizational Performance

Recall the previously mentioned book [6] that summarizes four decades of literature, through 2004, about how engineers communicate. The authors concluded that effective communication in engineering organizations increases personnel productivity, produces faster and higher-quality results, and saves money and other resources. Based on statistics for a wide variety of workplaces—going beyond just engineering organizations—effective communication increases productivity, improves team building and trust, and saves time and money [11]. Note the similar conclusions of these two sources.

1.3.3 Continued Learning and Self-Reflection

Some people seem to think that when a person writes or speaks about a topic, they are experts and, therefore, they simply write down or speak about what they already know. My experience and observations support a markedly different idea. One of the benefits of communicating, especially writing and speaking, is that both provide continued learning opportunities.

When I write, like me writing this book or preparing a conference presentation, I am studying and learning. Of course, I draw on what I already know, or think I know, about the topic. However, at the outset, I don't know what I don't know. Therefore, my preparation always reveals knowledge gaps, so I become a student to close them and to go even further. Sometimes the studying and learning part of writing or presentation preparation causes me to change my mind about something I was "sure of."

A paperweight in my office offers a thought from the French government official and poet Joseph Joubert, who over two centuries ago said, "To teach is to learn twice." That is, when you or I prepare to explain, via writing or speaking, something complicated to one or more interested individuals, we will naturally learn more about the topic.

Digging deeper, writing and presentation preparation are effective ways to clarify our thinking, to illuminate the more distant recesses of our minds. They become a powerful means of intrapersonal communication, that is, communication with ourselves. Thinking about our concerns, questions, ideas, theories, and options helps one "part" of us communicate with another "part" of us. Maybe it's our "brain" communicating with our "heart" or our "intuition" interacting with our "reason." Perhaps drafting text or preparing a presentation helps our left or logical brain communicate with our right or intuitive brain. Effective intrapersonal communication, achieved through writing or presentation preparation, can help each of us be more self-aware.

1.3.4 Affecting Change

Some noted engineer authors studied the multi-century history of the word "engineer," when used as a verb or noun. They concluded that, as a verb, engineer means to create, and when used as a noun, engineer means one who creates [12]. Therefore, if you are asked what you do as an engineer practitioner, or aspire to do after your student days, you could say, "create." If asked your occupation or planned occupation, you could say "creator." A practicing engineer could readily start the workday by saying, "What will I create today?" Briefly stated, creating is the essence of engineering.

Now let's temporarily shift emphasis to leading. Some of you are already leaders and others, but not necessarily all, will become leaders. I define a leader as someone who enables us, as individuals or more likely as a team, to accomplish great things, some of which we did not know we wanted to do. Leaders know how to envision better futures, assemble appropriate cognitively diverse teams, bring out the best in each team member, facilitate collaborative efforts, and produce "great things." They lead in a low-profile manner with the result being nicely defined by the Chinese philosopher Lao Tzu, who said, "But of a good leader, who talks little, when his work is done, his aim fulfilled, they will say, we did this ourselves."

Now, in one person, combine the idea that creating is the essence of engineering with leading. The result is the **engineer leader.** This person is rarely satisfied with the status quo and, therefore, frequently seeks great improvement or creative new approaches. The engineer leader knows how to affect change—how to deal with the inevitable initial, often knee jerk, opposition to change.

The engineer leader knows that converting opponents and neutral individuals to proponents requires persistence. Consider the following examples of now widely accepted and valued products [13–17]. Notice how some ideas were repeatedly rejected and, in other cases, how many years or attempts were required to move from concept to reality, from creative impulse to market.

- Electrostatic photography ("Xerox")—43 companies rejected the idea
- Dr. Seuss books—29 publishers showed no interest
- S. K. Rowling's first Harry Potter series book—12 publishers rejected it
- WD 40—40 attempts to get the correct composition
- Panama Canal—33 years from start of the first construction attempt until its opening
- Bar code—26 years from concept to first use, at a supermarket
- Golden Gate Bridge—20 years from concept to completed construction
- Velcro—10 years or more from concept to market
- Vulcanization of rubber—10 years to develop and patent the process
- Television—6 years to demonstrate a simple version

The engineer leader recognizes the need to persist and does it with effective communication. That leader uses various communication modes to gradually earn the support of opponents and neutrals. In my view, the most successful engineer leaders are competent communicators.

1.3.5 Personal Success and Significance

To the extent you continue to improve your technical and communication KSA and continue, by word and action, to hold public protection paramount, you will have increased opportunities to achieve success and significance within and outside of engineering. Consider success as that which benefits you and perhaps your family or other dependents. Success indicators include income, net worth, and material possessions. In contrast, view significance as your positive impact on others.

Success is about you, or me, and our "stuff," while significance is mostly about our positive impact on others. British Prime Minister Winston Churchill described the difference between success and significance by saying, "We make a living by what we get and we make a life by what we give." Most of us want both success and significance, but we vary widely in the preferred relative amount of each.

Leaders of and decision makers in various engineering organizations—academic, business, government, and volunteer—tend to value good to excellent communicators, especially those who are also technically competent. For example, three-fourths of practicing engineers participating in the previously mentioned North Carolina State University study rated writing as extremely important or very important in their performance. About 80% indicated that their performance reviews included oral and written communication [8]. Pair technical and communication KSA, and many success and significance doors will open for you.

1.4 COSTS OF POOR COMMUNICATION

The three examples presented at the beginning of this chapter begin to suggest the monetary and other costs caused by poor communication in or related to engineering projects. A more comprehensive list of such costs, along with examples, follows:

- Injury and loss of life—The Hyatt Regency Hotel and Boeing 737 MAX 8 disasters killed or injured hundreds of people.
- Monetary loss—Boeing paid 2.5 billion dollars in legal settlements, which included five hundred million dollars to compensate the families of the 346 people who perished.
- Damaged reputation—Examples are Boeing, Hyatt, and the engineering and construction firms involved in the Hyatt walkway collapse.
- Loss of clients—The municipality and engineering firm involved in the undocumented meeting case went their separate ways.
- License loss—The state of Missouri revoked the PE licenses of two engineers who worked on the Hyatt project.

- Loss of employment—Within nine months after the second Boeing 737 MAX 8 crash, the company board fired the Chief Executive Officer (CEO).
- Resignations—I learned about and communicated with two individuals who left Boeing prior to the two crashes because of their concerns with company engineering policies and procedures.

Let's conclude this discussion of the costs of poor communication with an anecdote. While it doesn't prove anything, this story reminds us, as individuals and organizations, to keep both technical and communication competencies in mind.

While serving as a consultant to an engineering firm, I attended some meetings of their executive team. During one of those gatherings, they discussed their frustrations with the "loss" of 17 clients. That is, 17 former clients no longer wanted the engineering firm's services.

The leaders turned to me as their consultant and asked what they could do. I suggested a direct approach. Contact, face-to-face or via telephone, each of the 17 clients. Ask each "lost" client to indicate what went wrong. The firm's leaders agreed to do that.

At the next meeting, those who contacted the "lost" clients reported these results:

- "Lost" because of technical problems: 2
- "Lost" because of poor communication: 15

In my view, stories like this happen too often, and hearing them reminds us that careless communication costs.

1.5 PRINCIPLES OF EFFECTIVE COMMUNICATION

The expression principles of effective communication means essentials applicable to the five communication modes presented in this book—asking, listening, writing, speaking, and use of visuals. The following six principles are widely applicable and, therefore, assumed and only briefly mentioned in the following chapters.

1.5.1 Know Your Audience

Assume you are composing an email to your counterpart at another engineering organization, about to call a representative of a potential employer to request a meeting, or are preparing a technical presentation to over a hundred attendees at an annual state engineering conference. Begin this communication effort by developing an audience profile.

Learn what you can about their positions, functions, responsibilities, disciplines, specialties, education, experience, goals, age, gender, ethnicity, and religion. Your audience will not be like you and some members may be markedly different. Get ready to speak to all of them, not you.

Weeks before a presentation to structural engineers, I asked the session organizer to help me understand the audience. His email reply provided this description: "The audience will be about five percent students, 30 to 40 percent owners, 50 percent designers and five percent other, including architects." That was helpful, a very diverse audience. I needed to speak to all members. You too can keep your audience's profile in mind—whether you will be communicating with one person or hundreds—as you select the content for your communication and decide how to present it.

Listen to the audience-oriented advice from a very effective speaker, Abraham Lincoln, United States' sixteenth President. He said, "When I get ready to talk to people, I spend two thirds of the time thinking what they want to hear and one third thinking about what I want to say."

Essentially, none of the documents we write will be "best sellers," as they say in the book publishing business. However, our documents—and our speeches—can "sell" ideas, solutions, services, and products if we know our audiences and tailor our writing and speaking to them.

1.5.2 State Your Purpose Clearly and Early

Articulate what you want the recipient of the email, the representative of the potential employer, or members of the conference audience to do. When I say articulate, I mean write one detailed sentence. The simple act of crafting that sentence forces you to define your purpose. Some examples follow. I want:

- the recipient of this email **to join** an engineering society committee that I chair and edit our reports
- the potential employer representative I am about to call to agree **to meet with me** next week about possible employment
- one or a few audience members **to contact me** and learn more about and possibly collaborate with me on my idea for reducing carbon dioxide emissions from construction projects

You may be thinking that this emphasis on profiling the audience and defining purpose sounds manipulative and controlling. It isn't, because I am urging you to be upfront and transparent about what you know about the audience and what you hope they will do. That approach makes effective and efficient use of their and your time and could lead to win-win results.

The recipient of the email you sent may join your committee and become a productive and happy editor member. The employer representative you call may decide to meet with you and then, during that meeting, decide that you are a perfect fit for a new position that the company is creating. One member of the conference audience, who owns a construction company, may offer to collaborate with you on the company's next project.

Start with where your audience is—not where you are—and try to move them toward a win-win outcome. That is what I am doing with this book. The Preface identifies two principal audiences, and Section 1.1.4 defines what I hope to accomplish with and for them, especially you.

Once you clearly define your purpose, plan to state it early in your document or presentation. I offer this advice for two reasons. First, it reduces audience anxiety about what is on your agenda and what you might ask them to do. Second, asking for something and asking early focuses the audience—it gets you and them aligned.

Your request does not have to be major. For example, I spoke to civil engineers about a sometimes-controversial American Society of Civil Engineers (ASCE) policy. It called for increasing the formal education of civil engineers, who want to eventually be in responsible charge of engineering projects, to a master's degree or equivalent. I asked audience members to try to understand why ASCE adopted this policy. I did not ask for agreement or support. That modest request seemed to reduce audience anxiety and enable them to focus on my message.

You could state your purpose early via an informative, nonthreatening title. For example, I used this title for a presentation at the annual conference of the Indiana Society of Professional Engineers: "Engineering Your Future in a Down Economy: 10 Tips." The title offers potentially useful ideas for consideration by audience members but does not ask for any action.

Sometimes we want audience assistance. If so, title the communication accordingly. I titled one of my books *Engineering's Public-Protection Predicament: Reform Education and Licensure for a Safer Society* [10]. A potential reader can expect to learn about a problem and be asked to assist with solving it.

Titles aside, and to reiterate, state your purpose clearly and early in your email, your telephone conversation, or your conference presentation. Especially with a conference presentation, tell them what you are going to tell them. Then tell them, that is, make your case. As you arrive near the end of the communication, concisely tell them what you told them. You could refer to the tell-tell-tell approach by calling it the T^3 method. We will expand on T^3 in Chapter 4.

1.5.3 Accommodate Their Preferred Ways of Understanding

Each of us tends to have a preferred way of understanding a message and later recalling and maybe using it. The person or persons who will receive your email, memorandum, letter, or brown bag presentation will likely prefer one of the following ways of receiving a message [18].

- **Auditory**—Understands primarily by hearing. Words are critical—visuals such as slides and props supplement understanding.
- **Visual**—Understands mainly by seeing. Relies heavily on slides, props, and other visuals.

- **Kinesthetic**—Understands mostly by touching, handling, and doing. When introduced to software, this person wants to get fingers on a keyboard and use the software.

Try to reflect sensitivity to the different preferred ways of understanding into your messages, regardless of the mode of presentation. This assumes that you know the recipient's preferences, which is often not the case. Consider some ways of dealing with known and unknown preferences.

You are composing an email to a colleague who you know has a visual preference. You ask for an opinion on the process you propose for conducting a new project. In addition to listing the sequence of steps, you attach a flowchart showing the steps, connections among them, and how some steps iterate back to earlier steps.

Assume you are preparing to speak to members of the local chapter of your professional society. You don't know nor can you control the types or understanding preferences that will be in your audience. Therefore, expect all three types to be present and do the following [18]:

- Thoroughly define terms and carefully select words for the auditory individuals. Later, in Chapter 4, we consider the importance of pre-presentation practice, with emphasis on what you say and how you say it, for the benefit of individuals who prefer auditory messages.
- Create visuals such as cartoons, drawings, graphs, icons, photographs, and videos for visually oriented individuals.
- Prepare handouts for audience members with kinesthetic and visual preferences. Make sure that they are stand-alone documents; that is, include the title of your presentation, date, event, sponsor, and your name, position, employer, email, telephone number, and website. These individuals tend to welcome handouts and take them to the office or their home. The contact information you provide encourages and enables post-presentation interaction between you and some audience members.
- Prepare props for the benefit of visual and kinesthetic individuals. The theatrical world uses the word "prop," a shortened form of property, which means any object used by an actor during a performance. Assume you are a mechanical engineer speaking about a new fabric for airbags in vehicles. You describe the fabric, show a slide of it, and then pass around a fabric sample, mainly for individuals with visual and kinesthetic preferences. See Section 5.6 for a detailed discussion of props, with examples.

The preceding advice for accommodating preferred ways of understanding when speaking to an audience may seem complicated. However, the next time you are preparing to speak to an audience, temporarily and perhaps humorously imagine a room full of ears, eyes, and fingers. Then begin to consider how you will connect with all three elements. Do this consistently, and you will begin to notice positive results over a series of presentations. Going forward, you are likely to incorporate, automatically, sensitivity to the different preferred ways of understanding into your presentations.

I occasionally observe the ramifications of trying to accommodate varied understanding preferences. For example, when asked to evaluate in writing one of my webinars, two responses were something like "creative, helpful graphics" and "silly, trivial graphics." Those markedly different responses suggested that I connected, respectively, with at least one visual individual and one verbal or kinesthetic person.

1.5.4 Recognize That What You Send Is Not Necessarily What They Receive

When you or I send a message, regardless of the mode (e.g., writing) or specific means (e.g., email), to someone or to a group, we sometimes intentionally or unintentionally make assumptions about the recipients. We assume that they will do some or even all of the following [19]:

- Understand the message
- Agree with the message
- Care about the message
- Act on the message

Whoever "they" are, we need to minimize relying on these assumptions. The communication mode discussions in the next four chapters include suggestions for confirming that recipients understand, agree, care, and/or will act.

1.5.5 Realize That You Are Surrounded by Communication Opportunities

If becoming an effective communicator is one of your goals, go way beyond waiting for a communication task assignment or being invited to communicate. With respect to the former, welcome a writing assignment from a professor, and later, when you enter practice, appreciate having a project manager ask you to draft some text for a project report.

Suggested Ways to Simultaneously Contribute Value and Gain Communication Experience

In addition to the communication experiences that automatically come your way, proactively seek ways to enhance your communication KSA during your formal education and later as you practice engineering. Communication opportunities abound within and outside of engineering. For example, you as a student, could:

- Take a for-credit speaking course
- Volunteer to be the writer, editor, or spokesperson for a group project in one of your courses

- Ask to take an independent study course in which you research an engineering topic of personal interest and then write a report and give a presentation
- Offer to speak about engineering to middle or high school students
- Write a letter to the editor of or an opinion article for the campus newspaper
- Tell selected faculty members about your communication interest and ask for their suggestions

Later, as a practicing engineer, you could:

- Agree to draft the minutes of your department's next meeting when the department head asks whether someone will take on that task
- Join a public-speaking improvement organization, such as Toastmasters International, a worldwide organization that "teaches public speaking and leadership skills through a worldwide network of clubs" [20]
- Indicate when a contentious issue arises on a project or within your office that you are willing to draft a clarification of the issue as the next step in the process of resolving it
- Offer, with your employer's permission, to speak about one of your projects at a local chapter of a professional society
- Contact local high school teachers and leaders and share your interest in speaking to students about engineering
- Write a letter to the editor of a newspaper or magazine (see Section 3.5.5) or contribute an opinion column (see Section 3.5.6)
- Tell your supervisor, colleagues, and friends about your communication interests and willingness to help with related tasks

Note that, with few exceptions, you can adopt many of the above-suggested actions without needing someone's approval and do so at minimal monetary cost. For that and other reasons, whether or not you become a good to excellent communicator is up to you.

What Would I Write or Speak About?

Assume you have successfully used some of the preceding ideas and have been selected to write or speak. What are you going to write or speak about? Sometimes others choose the topic because you have, via your efforts, become known as someone who is knowledgeable about a particular topic of interest.

Other times, you select the topic. A natural, initial thought is to select a topic about which you are an expert. However, another basis for topic selection is to choose a topic you would like to learn more about, consistent with the Section 1.3.3 discussion. This is one of the reasons I've enjoyed full-time and part-time teaching and writing throughout my career—helps me remain a student.

Another way to select a topic is to look objectively at the work you and your organization are doing. Might others be interested? Sometimes we are too close to what we do to see the value others may see.

As an example, a client and I gave two well-received presentations at national conferences on the role of mentoring in an engineering firm ownership transition. We simply described the process we used in his firm and shared the positive results. What's going on in your engineering college or engineering organization that might interest others?

Figure 1.2 illustrates the tactic of talking about what you and your organization do. Two costs appear on the horizontal axis. The first cost is whatever you already invested in one of your projects. That first cost might include labor, expenses, research, creativity, and energy required to complete a project for a client/stakeholder. The second or incremental cost is the additional cost of writing and/or speaking about the project.

The first and larger benefit on the vertical axis could include money earned, satisfaction, learning, and recognition resulting from the completed project. This is what we do day in and day out—we put something into our projects, and we and others get something out.

The short benefit on the vertical axis represents the additional benefit produced by writing and speaking about the project. This benefit might be improved speaking ability, exposure to your organization, network membership, and contributions to profession. It is larger than the additional effort expended.

You are likely to encounter Figure 1.2 situations in which the marginal cost of writing and/or speaking about some of your projects generates even larger

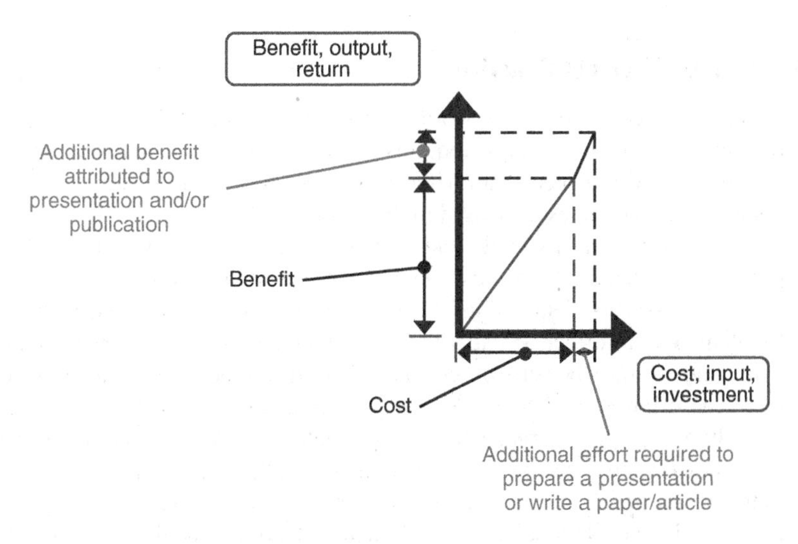

Figure 1.2 The small incremental cost required to write or speak about your work can yield a larger incremental benefit.

marginal benefits. And the benefit does not necessarily end there because of the possibility of leveraging the writing and/or speaking (as described in Section 4.6.4) and realizing additional benefits.

In summary, you do not necessarily have to create something so that you can write or speak. Instead, write or speak about the work you and your colleagues have done or are doing.

Personal Benefits of Proactively Seeking Communication Experiences

By proactively finding communication opportunities, you will earn and enjoy these benefits:

- Improved communication KSA, assuming you understand and use sound communication fundamentals as described in this book. Note the emphasis on understanding and practicing fundamentals in Section 1.5.6.
- Enhanced professional opportunities consistent with your values and goals. As stressed in Sections 1.2 and 1.3, communication is an essential part of successful engineering, and communicative engineers have earned an edge.

To reiterate a point made earlier, don't wait for an invitation to write, speak, or, in other ways, gain communication experience and make communication contributions. You probably won't live long enough to become a proficient writer or speaker. All engineers start out as amateur communicators and some remain so for their entire careers—amateurs rarely get invited. If you recognize and proactively act on the communication opportunities all around you, someday you will be an invitee.

1.5.6 Practice Perfect Practice

Achieving excellence in any worthy endeavor—cooking, golf, painting, piano playing, writing, speaking—requires understanding fundamentals and then selectively and repeatedly applying them. That observation applies to the five communication modes addressed in this book.

Geoff Colvin, in his book *Talent Is Overrated* [21], argues that "deliberate practice," not talent, defines the route to excellence in anything. I value the title of his book because many of us, when we see and/or hear excellence, attribute it to talent. We, in effect, say that the individual achieved excellence because of nature, not nurture, because of luck, not effort. Informed by my studies and experience, I attribute excellence to nurture, effort, and self-discipline.

Colvin explains, "Great performers isolate remarkably specific aspects of what they do and focus on just those things until they are improved; then it's on to the next aspect." He offers the example of Jerry Rice, one of the greatest pass receivers in the National Football League. Recognizing that he was not the fastest receiver, Rice committed to being one of the most difficult-to-tackle runners by defining and diligently practicing evasive moves. Tennis great Martina Navratilova may have been referring to deliberate practice when she said this: "Every great shot you hit, you've already hit a bunch of times in practice" [22].

Famed football coach Vince Lombardi described the same approach this way, "Practice does not make perfect. Only perfect practice makes perfect." At his first meeting with the Green Bay Packers during the 1961 training camp, he held up a football and said, "Gentlemen, this is a football" [23]. He first stressed understanding fundamentals and then focusing individual and team practice on ones where they were the weakest.

Call it what you want, such as "deliberate practice" or "perfect practice," focused practice clearly applies to communication. After learning the fundamentals of each of the five communication modes, we are likely to need focused practice on aspects of some of them to achieve excellence.

For example, you are about to speak to high school students about engineering. The last time you spoke to a class of high school students, you thought you had an opening statement that attracted attention. However, it did not seem to work. This time, you re-examine the content of the opening and refine it. Then, when practicing your speech aloud, you speak through the opening many more times than the remainder of the presentation. Having heard your opening many times and made appropriate changes in content and emphasis, you are confident it will grab attention. It does.

1.6 POISED TO BECOME GOOD TO GREAT COMMUNICATORS

I am fortunate to have worked in academia, government, and the consulting business. Therefore, I am familiar with the admirable academic and extracurricular activities of high school graduates who decide to study engineering—in contrast with students who select many other majors—and watched them succeed in challenging engineering studies. In government and business, I welcomed, supervised, and coached recent engineering graduates and other engineers as they joined our agency or firm. The vast majority proved quickly to be engaged and productive.

My observation: Engineering attracts the brightest and most persistent young people who, with high expectations and great support, will succeed. More specifically, they are poised to become good to great communicators because of their:

- Intellectual gifts, persistence, and other admirable characteristics
- Ability to be inspired by and learn from exemplar engineer communicators

Refer to Appendix B for brief introductions to some exemplary engineer communicators representing various engineering disciplines. Consider studying one or more of them.

Practicing good stewardship with the communication potential of future and current engineers requires helping them understand the importance of communication in engineering, teaching them communication fundamentals, and showing how to apply them. As indicated by its purpose, this book strives to do that.

1.7 KEY POINTS

- Communication means conveying, from one person to one or more others, ideas, information, and feelings, drawing on six modes—asking, listening, writing, speaking, visuals, and mathematics.
- Effective communication helps engineers meet their primary responsibility—public protection
- Other benefits of effective communication include improved organizational performance, affecting change, continued learning and self-reflection, and achieving personal success and significance
- Poor or miscommunication inevitably leads to monetary and other costs
- Principles of effective communication and advice about how to apply them enable engineers to use the multiple modes
- Engineering students and young practitioners are poised to become good to great communicators because of their intelligence and other gifts and because they can benefit from the examples of exemplar engineer communicators
- Each engineering student and practitioner should take the lead in becoming a communicative engineer

---

All the technology in the world
will not help us
if we are not able, at the core,
to communicate with each other and
build strong lasting relationships.

—Dorothy Leeds, communication consultant

REFERENCES

1. Fleddermann, C.B. (1999). *Engineering Ethics.* Upper Saddle River, NJ: Prentice Hall
2. Hoke, T. (2011). Ensuring the safety, health, and welfare of the public. *Civil Engineering.* ASCE. July.
3. Robison, P. (2021). *Flying Blind: The 737 MAX Tragedy and the Fall of Boeing.* New York: Doubleday.

4. National Society of Professional Engineers-Ethics. (2022). https://www.nspe.org/resources/ethics (accessed 20 December 2022).
5. Walesh, S.G. (2021). *Engineering's Public-Protection Predicament*. Chapter 4, Evolution of engineering's institutional framework: part 1-ethics and engineering societies. (pp. 132–138). Valparaiso, IN: Hannah Publishing.
6. Tenopir, C. and King, D.W. (2004). *Communication Patterns of Engineers*. (pp. 29–31, p. 91). Hoboken, NJ: Wiley and IEEE Press.
7. Ross, M. (2022). *The Bridge*. Resilient engineering identity. National Academy of Engineering. Spring issue. December 15.
8. Swarts, J., Pigg, S., Larsen, J., et al. (2018). Communication in the workplace: What can NC State Students Expect? North Carolina State University. Professional Writing Program.
9. National Society of Professional Engineers. (2013). *Engineering Body of Knowledge-First Edition*. Alexandria, VA: NSPE.
10. Walesh, S.G. (2021). *Engineering's Public-Protection Predicament*. Chapter 3, Disasters: Were some caused by licensure-exemption cultures? Valparaiso, IN: Hannah Publishing.
11. Barraclough, D. (2023). The importance of effective workplace communication—statistics. Expert Market. https://www.expertmarket.com/phone-systems/workplace-communication-statistics (accessed 20 July 2023).
12. Walesh, S.G. (2012). *Engineering Your Future: The Professional Practice of Engineering*. Chapter 1, Introduction: engineering and the engineer. (pp. 283–285). Hoboken, NJ: Wiley.
13. Walesh, S.G. (2017). *Introduction to Creativity and Innovation for Engineers*. Chapter 6, Characteristics of creative and innovative individuals. Boston, MA: Pearson Education.
14. Walesh, S.G. (2012). *Engineering Your Future: The Professional Practice of Engineering*. Chapter 15, The future and you. (pp. 442–446). Hoboken, NJ: Wiley.
15. Madrigal, A.C. (2011). Why WD-40 is called WD-40. *The Atlantic*. September 27. https://www.theatlantic.com/technology/archive/2011/09/why-wd-40-is-called-wd-40/245734/ (accessed 13 May 2023).
16. Britannica. (2023). Panama Canal. https://www.britannica.com/topic/Panama-Canal (accessed 13 May 2023).
17. Wikipedia. (2023). J. K. Rowling. https://en.wikipedia.org/wiki/J._K._Rowling (accessed 13 May 2023).
18. Walesh, S.G. (2012). *Engineering Your Future: The Professional Practice of Engineering*. Chapter 3, Communicating to make things happen. (p. 102). Hoboken, NJ: Wiley.
19. Clarke, B. and Crossland, E. (2002). *The Leader's Voice*. New York: Select Books.
20. Toastmasters International. (2022). https://www.toastmasters.org/about (accessed 27 December 2022).
21. Colvin, G. (2008). *Talent Is Overrated: What Really Separates World-Class Performers from Everybody Else*. (pp. 52–56). New York: Penguin Group.
22. Verma, G. (2022). He's kicking butt' – Martina Navratilova once heaped praise on a young Tiger Woods after scripting back-to-back victories. Essentially Sports. https://www.essentiallysports.com/wta-tennis-news-hes-kicking-butt-martina-navratilova-once-heaped-praise-on-a-young-tiger-woods-after-scripting-back-to-back-victories/ (accessed 6 March 2023)
23. Maraniss, D. (1999). *When Pride Still Mattered: The Life of Vince Lombardi*. (p. 274). New York: Simon & Schuster.

EXERCISES

Notes to instructors:

- **a.** I assume instructors will use this book in at least two different ways. It could be the textbook for a communication course or a supplementary book for use in many courses, each of which includes some communication instruction.
- **b.** This, and the other chapters, present exercises as though individuals will complete them. However, consider assigning some exercises to small teams (e.g., three students). Working together as teams on exercises helps students hone their communication and teamwork KSA.
- **c.** I consider the exercises at the end of each chapter to be a get-started list for instructors and students. Many instructors prefer to create many of their own exercises based on their teaching and practice experience and what they know about their students.

1.1 Role of communication in disasters: Investigate the January 28, 1986, space shuttle *Challenger* disaster. Determine the role of communication in that tragedy. Describe your findings.

1.2 Role of communication in effecting change: Choose one of the creative results in Section 1.3.4, or select some other creative result that interests you. Give an example of how effective communication assisted implementation or how poor communication frustrated achievement.

1.3 Preferred way of understanding: What is your preferred way of understanding? Give an example of what works for you, and explain why.

1.4 Learning from exemplar engineering communicators: Select one of the engineers featured in Appendix B. Find something they wrote or a speech they gave, in addition to the quote offered in the appendix. Indicate who you selected and rate that person's communication effectiveness on a scale of 1–5 (very ineffective, ineffective, neutral, effective, very effective). Explain your rating.

1.5 Book review—a multi-month exercise: This exercise will introduce you further to communication literature by asking you to read and report, in writing and/or speaking, about a nonfiction communication book. Do the following:

- **a.** Select a book that addresses some aspect of communication. It could focus on one mode of communication, such as speaking, or take a more comprehensive approach. Your selection might be about communication in engineering or have nothing specifically to do with engineering. To get started, consider skimming the cited references listed at the end of the chapters in this book. Then broaden your search.

b. Obtain approval of the book from your instructor and determine whether you are to produce a written review, spoken review, or both.

c. Read the book while starting to write a review and/or prepare a speech review of it. Include the following topics in your review:

- Title, publisher, year published, and author
- Principal message or messages, that is, what is the author trying to accomplish?
- Your views—pro and con—of the book including how well the book achieves the author's purpose
- How the book will, or will not, help you become an even better communicator
- Submit the written review and/or provide a spoken version of your review

1.6 Research project—a full semester/quarter exercise: This major exercise has these three objectives:

a. Provide you with added knowledge and skills needed to take on a research, writing, and/or speaking project—learning by doing.

b. Explicitly connect engineering and communication—successful engineers tend to be effective communicators and successful engineering projects tend to include effective communication.

c. Show you how engineering projects typically have major technical and nontechnical elements.

Do the following:

a. Based on your interests, select a completed engineered product, facility, structure, system, or process that you want to learn much more about.

b. Request approval of the topic from your instructor and determine whether you are to produce a written report, spoken review, or both.

c. Conduct your research and structure your report or spoken presentation using all of these major headings: milestones, technical challenges, legal requirements, environmental issues, safety provisions, financing, proponents, opponents, and lessons learned. Although the headings are required, they don't need to appear in the indicated order.

d. Because this exercise will be underway throughout the semester or quarter, continuously look for ways to incorporate what you are learning about communication, from this book and in class, into your evolving report or spoken presentation.

CHAPTER 2

ASKING AND LISTENING

My greatest strength as a consultant
is to be ignorant and ask a few questions.

—Peter Drucker, management consultant and writer

After studying this chapter, you will be able to:

- State some of the benefits of the asking and listening process
- Explain obstacles to effective question-asking and identify the major one
- Describe five question-asking methods
- Discuss empathetic listening and the role of body language in achieving it
- Provide advice for effective asking and listening

2.1 BENEFITS OF QUESTIONS

2.1.1 The Problem: Valuing Being Understood More than Understanding

Imagine two people or several individuals—in both cases including some engineers—beginning to discuss a mutual concern. Maybe it is defining a technical problem, resolving a personnel issue, or pursuing a business opportunity. You are one of the participants in this face-to-face (F2F) conversation.

The discussion could go quickly in various directions, many of which are not productive. For example, one person could dominate the conversation by delivering a monologue or, in some other way, forcefully express his or her views to the virtual exclusion of others. Another participant might say that the solution is "obvious" and go on to dictate, "What we should do." Someone may repeatedly interrupt others in mid-sentence. The preceding miscommunications can lead to

The Communicative Engineer: How to Ask, Listen, Write, Speak, and Use Visuals, First Edition.
Stuart G. Walesh. © 2024 John Wiley & Sons, Inc. Published 2024 by John Wiley & Sons, Inc.

personal frustration; failure to correctly define the problem, issue, or opportunity; "solving" the wrong problem; and ongoing interpersonal conflict.

Desiring to be understood, not to understand, is the overriding flaw in the preceding scenario; the need to share views with others while not wanting to learn their views. To succeed, communication must be a two-way street.

2.1.2 The Solution: Asking and Listening

Resolution of this miscommunication challenge lies partly in recognizing the value of routinely asking questions and then carefully listening to the answers. When you or I ask a question, we are in effect saying, I know, or think I know, something about this topic. However, I want to know what you think—hear your view. The question-asking and subsequent listening are powerful because both askers and listeners benefit; they learn from each other. Now we have that desirable two-way street.

More specifically, the questions that begin the mutually beneficial asking–listening process stimulate and enable the following five communication benefits [1, 2]:

1. **Questions create an obligation to respond:** Recall, in an engineering class, when a professor asked you a question and then was silent. You were naturally inclined to fill that silence, and, therefore, you tried to answer the question or perhaps deflect it by requesting clarification. I am not suggesting that we use questions to cause embarrassment or discomfort. Instead, politely and diplomatically ask a question and then create some silence to provide the other person an opportunity to frame and offer an answer.
2. **Questions stimulate the thinking of both the asker and answerer:** You, as the asker of a series of questions, must think broader and deeper about the overall topic before articulating your next question. The person you are interacting with will be thinking in a similar manner as they listen to and prepare to respond to your last question. Thinking stimulated by question asking and answering leads to thoroughly defining a problem, issue, or opportunity and then finding an effective course of action.
3. **Questions cause sharing of valuable data, information, and knowledge:** Your thoughtful questions indicate your concern for the other person and reveal your interest in the topic under discussion while, at the same time, providing useful data, information, and knowledge to the other person. Answers to your questions enable you to learn more about that individual and provide you with the data, information, and knowledge you can use to contribute more effectively to joint efforts.
4. **Questions temporarily put the asker "in the driver's seat":** A proactive and thoughtful question asker can use questions to focus a discussion or direct it in a more productive direction, all in the interest of producing mutually beneficial results. Of course, a responder could use an answer or a counter-question to steer the discussion in another direction.

5. **Questions enable people to persuade themselves:** Thoughtful asking and listening, in contrast with autocratic directives, tends to move all conversing participants toward a common understanding and resolution. According to electrical engineer, Charles F. Kettering, "a problem well-defined is a problem half-solved." Building on the foundation of a collectively defined challenge, continued asking and listening will lead the group to an optimum resolution.

In my view, the smartest person in the room is frequently the one who understands and values the preceding benefits of the asking–listening process. Accordingly, that person often talks the least and, when speaking, frequently asks questions. The late Malcolm S. Forbes, an American entrepreneur and publisher of *Forbes* magazine, supported that view. He added an interesting nuance when he said, "The smart ones ask when they don't know. And sometimes when they do."

At the other end of the talking spectrum, we find those who talk too much. Author Dan Lyons refers to these individuals as overtalkers and says they suffer from the addiction of talkaholism. Lyons suffered from talkaholism, developed and used guidelines to greatly reduce his overtalking, listened more and more effectively, and improved relationships. He claims that the most powerful and successful people talk little and listen a lot [3].

My goal in writing this chapter is to optimize the results of conversations among two or more individuals. I do this by sharing question-asking and listening knowledge, skills, and attitudes (KSA) and inviting you to apply them. This chapter is about improving interpersonal communication, not about manipulation, deceit, or tricking others to reveal confidential information.

2.2 QUESTION-ASKING OBSTACLES

Given the desire to balance our need to be understood by others with our need to understand others, and given the five ways questions do that and in other ways improve communication, one would expect we engineers to be exuberant question askers. Not so, based on my experience and on polls I've conducted during webinars [4, 5]. While we may recognize the value of questions, we also encounter, or think we encounter, barriers to asking them. Accordingly, we do not ask enough or the most probing kinds of questions. As an aside, this is not a problem unique to engineers—I suspect it prevails in many occupations and professions.

I anonymously polled participants of webinars and asked them to select their principal obstacle to asking questions based on the following:

1. Reluctance to question authority
2. Coming across as intrusive or rude
3. Appearing uninformed or poorly prepared
4. Other

The most common obstacle, by far, was number three. Having said that, consider the logic, or lack thereof, of each barrier.

1. **Reluctance to question authority:** Recognize that most of us are authorities, just on different things. The person you interact with is an authority on some things. Similarly, you, as a professional, or aspiring professional, are an authority on other things. By asking questions, you are not questioning the authority of the other person. Instead, you are reflecting and sharing yours and trying to benefit from theirs.
2. **Coming across as intrusive or rude:** Someone said, "I don't care how much you know, until I know how much you care." We demonstrate care, not rudeness, by preparing and asking thoughtful, probing questions. Of course, we pose them in a polite, sensitive manner.
3. **Appearing uninformed or poorly prepared:** As illogical as it may seem, too many engineers fear that asking questions will cause others to view them as not being knowledgeable or ready. Asking questions does not indicate you are uninformed or poorly prepared. It should mean just the opposite, that is, because you are well informed and thoroughly prepared, you know what to ask. The type and number of questions you ask reveal your expertise—and your care.

   Consider this hypothetical scenario: Tomorrow morning, you awake with a pain in your chest and quickly arrive at the hospital emergency room. The doctor asks, "What's wrong?" You answer, "Chest pain," and the doctor, based on your comment, says, "We are immediately performing triple by-pass heart surgery." The pain aside, how would you feel? Might you want the doctor to ask more questions as part of a careful diagnosis of your problem before deciding how to solve the problem? Maybe something you ate caused the pain.

   Now consider an actual question-asking situation that illustrates the value, to all parties, of preparation [2]. I was preparing for an interview with leaders of an architectural and engineering firm, regarding a potential education and training project. I prepared a list of 24 questions for the potential client with the hope that I would be able to ask a few of them. Soon after meeting the leaders, I began to think that they had not prepared very well, which their spokesperson confirmed by saying something like, "So what would you like to know?" I asked and they answered essentially all of the questions and, before I left, they retained me to do the project. I believe my questions indicated that I cared, was informed, and was prepared.
4. **Other:** Webinar participants' main concern here was not knowing what questions to ask and/or how to ask them. Appendix C provides example questions for various common situations. The next section describes ways to ask questions.

2.3 QUESTION-ASKING METHODS

Think of the five question-asking methods described in this section as a toolbox from which you select one or more tools to use in a given situation. They will help you and those you interact with realize the communication benefits described in Section 2.1.2. By judiciously using these question-asking methods, you will "drill down"—get to the bottom of things—and move past symptoms to discover causes.

For example, you might begin a discussion with closed-ended and open-ended questions, use some of Kipling's six, and end with the five-why analysis. You go up the learning curve and gain a better understanding of a problem, issue, or opportunity and the other person.

2.3.1 Closed-Ended and Open-Ended Questions

Closed-ended questions typically have short answers, such as yes, no, a fact, a number, or a name. Consider some examples:

- How much is budgeted for the electronics portion of the project?
- Where will project implementation funds come from?
- Who will be the principal obstacle to project success?
- What data/information are most suspect?

You could easily imagine a one- or several-word answer to each of the preceding questions.

Open-ended questions, in contrast with closed-ended ones, usually have longer answers and give or invite the recipient an opportunity to reflect, explain, and elaborate. Some examples:

- Why are you planning to modify the existing bridge rather than construct a new one?
- What else should I know so that I can assist you?
- Why are you assuming that a spread footing is the best foundation for the transmission towers?
- When this project is completed, how will you and others determine how well it succeeded?

Because they generate so much information and raise so many issues, open-ended questions easily lead to more closed and open-ended questions. If you, as a student or practitioner, want to know the facts and feelings behind any project, use a combination of closed and open-ended questions.

The last example of open-ended questions warrants additional comment. The success of most engineering projects will include the degree to which they satisfy schedule, budget, and deliverable requirements—the traditional basics. However, never assume that those are the only, or even the most important, success indicators. The person or persons you serve with engineering, as a member of a consulting firm serving a client, or as an employee working on a project assigned by a supervisor, may have other interests. Some examples: presenting excellent project results to support a request for assignment to bigger projects, using the completed project as part of the proposal to obtain a grant for another project, authoring an article about the project, or looking good within their peer group.

Another way to think about success indicators is to heed the advice of Stephen Covey, author of *The Seven Habits of Highly Effective People* [6]. His Habit 2 is "Begin with the End in Mind." In similar fashion, the Greek philosopher Plato stated, "The beginning is the most important part of the work."

Separated by two and a half millennia, these two advisors urge us to determine, up front, what a successful project or other endeavor would look and feel like. Ask various stakeholders open-ended questions that probe how project success will be determined—listen for facts and emotions. Articulate in writing that end point, then go back to the beginning, plan, and conduct the project to achieve it.

Engineers and other technical and scientific personnel tend to ask more closed-ended than open-ended questions—we value facts. This is good, but not sufficient, if we want a broad and deep understanding of a challenging situation. Seek facts plus the related motivations and emotions.

A response from one of the two types of questions can lead to asking a question of the other type. For example, you learn, in response to a closed-ended question, that the research and development (R&D) department in your company is likely, based on experience, to be the principal obstacle to the innovative technical approach you are proposing. This might prompt you to ask this open-ended question: Why has R&D usually opposed this kind of project?

Maybe question-asking, whether closed or open-ended, isn't your forte, or you simply feel uncomfortable doing it, especially in professional matters. Perhaps you are dealing with some of those question-asking obstacles discussed in Section 2.2. Then experiment with asking more questions and listening in low-risk situations.

Consider saying something like this to the cashier at a restaurant: "I like the new menu—why did you make the changes?" Or, in a first-year engineering class, after hearing how to create and use a free-body diagram, ask the instructor why we need to know this? Consider attending a meeting of an organization that is new to you, such as the student chapter of an engineering society or the local Chamber of Commerce, and learn what you can by asking questions. By experimenting and practicing, you will increase your knowledge—the answers to your questions—and become increasingly comfortable with asking and listening. It could become a productive habit.

2.3.2 Kipling's Six

Rudyard Kipling, the English writer and journalist, offers another question-asking method for your consideration. His 1902 poem, "I Keep Six Honest Serving-Men," included the following: "I keep six honest serving-men, they taught me all I knew. Their names are What and Why and When and How and Where and Who" [7].

Kipling's six helpers and teachers can immediately assist you in many and varied conversation situations. Imagine that you, as an engineering student or practitioner, attend the monthly dinner meeting of your local chapter of the National Society of Professional Engineers (NSPE). You are seated at a round table with four other individuals, none of whom you know. One of the others mentions a new technology—let's call it NEW. You know nothing about NEW, want to learn about it, and want to contribute to the conversation. What do you do?

You use Kipling's six. During the course of the half-hour discussion, you unobtrusively, politely, and at appropriate moments ask the questions listed below or something similar.

- I am not familiar with NEW; **what** is its purpose?
- **Why** was it developed by the private sector?
- **When** did that development process begin?
- **How** can I get current articles, papers, or books describing it?
- **Where** is the headquarters for the organization that developed and supplied it?
- **Who is** the most knowledgeable NEW expert?

By the end of the conversation, you have learned about NEW, contributed to the learning of others, and possibly made some potentially beneficial personal contacts—all because of questions stimulated by Rudyard Kipling's six.

At some point in your career, you will, intentionally or unintentionally, meet with a potential client, owner, customer, or stakeholder of the organization—business, government, academic—that you work for. Such interactions, if they begin well, sometimes eventually lead to mutually beneficial results. You can start that win-win process by applying some of Kipling's six. Because Kipling urges you to ask questions, you reduce the likelihood of talking too much about yourself and increase the likelihood of learning about the other person. While your focus on others, via questions, helps others understand you, it also enables you, via their answers, to understand them.

2.3.3 The Five-Why Analysis

Assume you are a consulting engineer meeting with a potential new client, the manager of a large manufacturing plant, who has expressed interest in having more or improved entrances and/or exits on the employees' parking lot. You apply the five-why analysis [8, 9] as follows:

1. Why do you need more entrances/exits for your parking lot? ANS: Because the lot entrances/exits cannot handle the growing number of vehicles. (You could stop here or at subsequent points in the process, but

you don't. You politely press on until you have exhausted what appear to be all possible causes of the parking lot problem and/or have gotten to the root cause).

2. Why does the lot fail to accommodate the number of vehicles? ANS: We have major congestion and related frustration when workers enter and leave the lot. (You could stop here and concur with the need for more or better entrances/exits—but you don't).
3. Why does this congestion occur? ANS: We operate five days per week using three 8-hour shifts. On those days, first-shift workers leave the lot at 4:00 p.m. when the second-shift workers arrive. The same volume of entering and leaving occurs at midnight and 8:00 a.m.
4. Why do all the shift workers arrive and leave at the same time? ANS: Because we have always done it that way. (In any questioning scenario, be aware of this possible "we've always" situation and, if you discover it, probe more in that direction.)
5. Why do all workers in a shift need to start and stop—arrive and leave—at the same time? ANS: They don't—we could stagger start and stop times in, say, 30-minute intervals. That would solve our problem and save us engineering and construction fees. I like the way you help us think through a problem.

The example five-why analysis illustrates how the method can get to the root cause of a problem. Which, as noted in Section 2.1.2, renders the problem "half-solved."

The preceding example ends with the potential client apparently not needing your and your firm's services. In effect, you used the five-why method to talk yourself out of a project. I ended the five-why story that way so that I could alert you to an inevitable predicament faced by consulting engineers and their firms.

Let's explain the predicament by asking this question: What is your higher priority as a provider of professional services, getting that first project with a new client, or starting to develop a mutually trustful relationship? In the parking lot scenario, you could have gone in a different self-serving direction to win that first contract. But you did not—you acted in the best interest of the potential client by asking a series of questions that led to the root cause of the problem and the solution to it. You probably took the first step in building a long-term, win–win, client–consultant relationship.

Toyota makes widespread use of the five-why approach. Consider an example taken from their problem-solving education and training program [8]. Someone sees oil on the floor beneath a machine, asks why, and is told the machine leaks oil. So why is it leaking? Because a gasket deteriorated. Why did it deteriorate? Because we buy inferior gaskets. Why do we buy poor gaskets? Because they are cheap. Why do we prioritize buying a cheap product and getting only short-term results? Because our purchasing agent is rewarded for short-term cost savings.

Notice how, in this example and in most applications, the process could stop whenever asking "why" finds an implementable solution, although it might not be the best solution. In general, the more often we ask "why?" the more likely we will discover the root cause. The root cause of the oil-on-the-floor problem was the means used to reward purchasing agents. On seeing the oil, someone would have to be a clairvoyant genius to see the root cause. The good news: You or I don't have to be geniuses to first avoid jumping to conclusions about the root cause of a problem and, second, apply the five-why analysis to find the root cause.

A potential downside of the five-why approach is that we miss the concept of asking a series of probing questions and, instead, fall into a mechanical, if not rude, why-why-why-why-why pattern. Don't interrogate; instead, finesse the questions [10]. For example, the first question in the parking lot story could be softer, such as "Please help me understand why you need more entrances/exits for your parking lot."

The "5" in the five-why analysis is not rigid. It is large enough to motivate us to be thoughtful and thorough, to think broad and deeply in our questioning, and could be exceeded.

2.3.4 The Echo Technique

When listening to another person, you can use this handy method to either confirm your attention to and understanding of what he or she is saying or to, in effect, ask a question [11]. Imagine you are on the staff of an engineering firm that helps industry clients improve the effectiveness of their manufacturing processes. You, an expert, are meeting at a client's place of business for the first time with the manager of several manufacturing lines. After a brief introductory discussion about a persistent problem, the manager says, "Line B has shut down six times this week."

You, diplomatically and in echo manner, repeat the manager's last four words (or some variation on them) in one of the following two ways:

- ". . . six times this week" expressed in a calm, matter-of-fact tone.
- ". . . six times this week?" expressed in a surprised and very concerned manner and as a question.

Both short comments confirm that you are paying attention and are concerned. The second comment suggests urgency—you are eager to learn more, such as what part of the line failed, for how long, and at what cost, and then help the manager solve the shutdown problem. You use the echo technique to set the tone for what you hope will be a productive, win–win relationship with the manager.

I mentioned that you are an expert. You have manufacturing line expertise. You also have communication expertise, including how to carefully repeat the words of others to confirm that you are listening, understand, and intend to use what you learn to help them.

2.3.5 The What If? Method

When faced with an especially challenging problem, issue, or opportunity, we are likely to seek a questioning approach that stimulates fresh, whole-brain, and creative thinking. The what if? method does that.

Before adding this tool to your question-asking toolbox, you may initially view it as redundant. For example, the "what" in what if? appears to be part of the closed-ended and open-ended questions tool and Kipling's six method. Yes, but the surprising and stimulating words that follow "what if?" distinguish and empower the method.

Consider three examples of the apparent application of what if? thinking. The challenge of finding such examples is that, while we recognize the creative results and appreciate them being stimulated by initially unrestrained thinking, we cannot know what the creators actually thought, let alone know that one of them said, what if we. . .?

Rebuilding a Taco Bell Restaurant

In August 1965, riots in the Watts area of Los Angeles caused 34 deaths, over 1000 injuries, and 40 million dollars in property damage. That damage included the destruction of one Taco Bell restaurant. The company committed to rebuilding the restaurant. In the spirit of what if?, they vowed to do the rebuild within a 48-hour period as a means of publicizing their intent to rebuild. This was a bizarre idea in that restaurant construction typically took months. Via very careful planning, they completed construction in just under 48 hours, earned positive publicity, and learned how to reduce the cost of future conventionally constructed restaurants by 20–50% [9].

Completing the Panama Canal

In 1905, engineer John F. Stevens took over leading the construction of the struggling 51-mile-long Panama Canal, which predecessors viewed as an excavation project—many steam shovels were available but hauling away the excavated material was a bigger challenge.

Using what in retrospect appears to be what if? thinking, Stevens freshly and creatively viewed the canal effort as a railroad project, not an excavation project, and thus began the Railroad Era. They designed and built a system that kept an empty or filling railroad car next to every steam shovel all the time during every work shift. The system included machines that picked up and moved track sections so that rails and cars were always close to the steam shovels. With the principal obstacle removed, the project moved forward, and the canal opened in 1914 [9].

Temporarily Storing Potential Flood Water on Streets

Back in the 1980s, Skokie, IL, a densely developed 8.6 square mile suburb north of Chicago, increasingly faced severe basement flooding because of backup of sewage from combined sewers—carry wastewater and stormwater—during even moderate rainfall events. Recognizing that out-of-control stormwater was the problem's cause, engineers asked what if they designed a system that temporarily stored stormwater on the street, to prevent combined sewer surcharging and basement flooding, and then slowly released the water into the combined sewers. Although the idea was radical at that time, the system was designed, constructed at a cost that was 38% of the cost of the next best option and solved the problem [9].

Let's conclude this description of what if? by reminding ourselves, as illustrated by the examples, that this thinking tool stimulates fresh, whole-brain, and creative thinking. However, some engineers and others may be reluctant to propose and discuss what appear be bizarre ideas for solving real problems. If you encounter that resistance, share some of the above stories with them.

2.4 LISTENING

Up to this point, this asking and listening chapter focuses on the former by discussing the benefits of questions, obstacles to question-asking, and question-asking methods. We ask questions to generate information, ideas, and feelings that we can use to resolve problems and issues or pursue opportunities. Therefore, we need to listen carefully to answers to our questions, so we hear what is said and not said but conveyed. "The most important thing in communication," according to consultant Peter Drucker, "is to hear what is not being said."

Of the five modes of communication discussed in this book, listening might appear to be the easiest. Questioning, writing, speaking, and using visuals all seem to require more effort than listening. In contrast, listening just automatically happens. No, hearing just happens; listening requires effort.

2.4.1 Listening Is Much More Than Hearing

Hearing is "the process, function, or power of perceiving sound... the special sense by which noises and tones are received as stimuli" [12]. As explained by author Linver [13], hearing "is a natural, passive, involuntary activity. Anyone with a normally functioning ear and brain will involuntarily hear sounds of certain intensity."

While listening requires hearing, it goes way beyond that to include understanding what others mean and how they feel. Effective listening means being attentive, verifying understanding, and using what we learn. American philosopher, author, and educator Mortimer J. Adler explained the sharp distinction between listening and hearing as follows: "Listening, like reading, is primarily an activity of the mind, not the ear or the eye. When the mind is not actively involved in the process, it should be called hearing, not listening; seeing, not reading."

2.4.2 Listen and Look for Both Facts and Feelings

Engineer and author Samuel C. Florman, PE, offered this critical observation: "One of the failings of engineers is they overestimate the power of logic and underestimate the power of emotion." We might paraphrase him by saying that one of the communication shortcomings of some engineers is that, when listening, they overestimate the power of getting the facts and underestimate the power of learning feelings.

Engineers eventually learn that fully understanding a problem, issue, or opportunity, no matter its magnitude, requires getting the facts and just as certainly discerning the feelings of the individuals who have a stake in the matter. Facts and feelings, logic and emotions typically drive the decisions we make and actions we take in academia, business, government, and other settings [14].

As illustrated in Figure 2.1, we can recognize five levels or steps of listening [6]. I suspect that each of us has experienced every one of them, but maybe not in the context of all of them. If we strive to at least occasionally step up to empathetic listening, our interactions with others will improve.

Consider these brief descriptions of the five listening levels starting with ignoring:

- Ignoring—we hear someone talking but give no attention to what they are saying.
- Pretending to listen—we pretend to listen, in the interest of appearing to be polite, but we do not engage; our thoughts are elsewhere.
- Selective listening—we hear, and try to understand, only what we agree with or otherwise want to hear.
- Attentive listening—we hear and try to comprehend everything, that is, facts, information, and ideas.
- Empathetic listening—we strive, in addition to grasping facts, information, and ideas, to understand how the speaker feels.

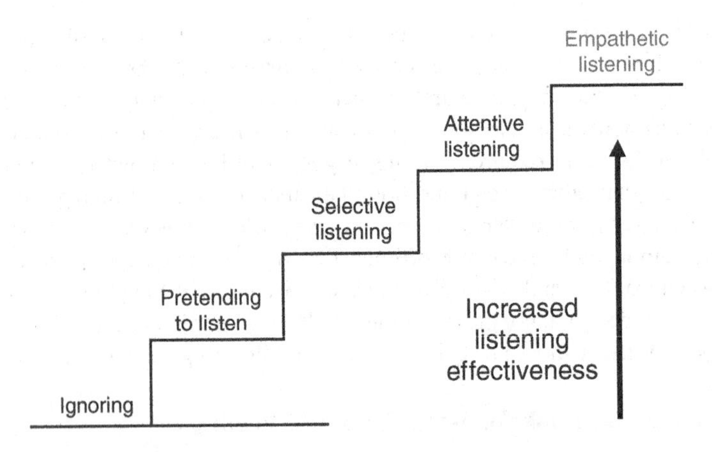

Figure 2.1 Listening occurs at five levels, the most effective of which is empathetic listening.

According to author Amy E. Herman, who specializes in observation and communication, "the ability to imagine others' viewpoints, reactions, and concerns is one of the most important cognitive tools we humans possess" [15].

While empathetic listening is the highest and most effective level you or I can achieve, it does not necessarily mean sympathetic listening. Our understanding of someone's feelings about a problem, issue, or opportunity is not the same as our sharing those feelings. However, this in no way detracts from appreciating their views because, going forward, that understanding helps us relate to and interact with them—and, in some circumstances, we might alter our thinking and feelings.

Consider a personal example of the value of knowing about feelings. My firm assigned me to manage an innovative and potentially large project for a municipality. I was excited because we—the community and the firm—had an opportunity to demonstrate a new way to manage stormwater runoff. By asking and listening, I learned that the municipality's chief engineer, who was my principal contact, did not share my excitement. He valued the innovative aspects of the project but had negative feelings about the joint effort because he wanted to work with another consulting engineering firm—not mine.

Forearmed about his feelings, I did what I could to earn his respect and, to his credit, he made a similar effort. We gradually developed a good relationship, and the multiyear project performed as expected. As another indication of our success, he and I wrote and published an article about the project.

After urging us to strive for empathetic listening, Stephen Covey notes that it is "risky" [6]. If we successfully use empathetic listening, we will learn how another person feels about some topic. On recognizing that their feelings are different than ours, and hearing why, reason and conscience may compel us to change how we feel. We could begin a conversation confident in what we know and how we feel and end it knowing more, as expected, and feeling different, which was not expected.

Besides listening for indications of emotions, we can also use our eyes to look for signs of emotions, which leads to our next topic, body language. The communicative engineer gradually develops the ability to listen and look for facts and feelings.

2.4.3 Body Language

Six decades ago, anthropologist Edward T. Hall used the expression "the silent language" for what we now usually call body language, that is, nonverbal communication such as posture, facial expression, arm positions, handshake, eye contact, and dress. Hall said that silent or body language, functioning "in juxtaposition to words," conveys feelings, attitudes, reactions, and judgments" [16]. Bodily actions and interpretations of them are likely to vary across cultures.

Schoolmaster and minister, Ralph Waldo Emerson, stressed the importance of body language by offering this advice: "When the eyes say one thing and the tongue another, a practiced [person] relies on the language of the first" [17]. Accordingly, if you or I want to be an effective listener, especially when understanding emotions, we should "listen" with our ears and our eyes. Perhaps actions speak louder than words when we are trying to understand the feelings, attitudes, reactions, and judgments of others.

Examples of body language exhibited by someone listening to you [15, 18, 19]:

- Arms crossed on chest—resists your message, closed mind
- Absence of eye contact—does not want to engage in conversation
- Touching nose—thinks your words are deceptive
- Hand on back of neck—has questions/concerns
- Raised eyebrow(s)—does not believe you
- Looking at ceiling—deciding or maybe bored
- Relaxed and smiling expression, eye contact—agrees

As you observe and interpret body language, seek consistency and repetition. One type of body language shown for an instant is likely to have little meaning. See if it repeats and look for its consistency with other body language messages. The person you are communicating with may temporarily cross their arms against their chest because they are cold, touch their nose because it itches, or look at the ceiling because they admire the fan [14].

As important as feelings may be, the person you are conversing with and trying to get to know better may initially be reluctant to share feelings, even if you ask. However, by consistently asking many and varied questions and listening with your ears and eyes, you will gradually discover some feelings. Some possibilities: Pride in department, frustration with declining profitability, reluctance to change, fear of losing job, open to new ideas, or dealing with a nonwork challenge.

You are likely to travel internationally or work with clients, customers, and colleagues around the globe. Expect the meaning of nonverbal communication to vary among countries. For example, when speaking to a Japanese audience, don't expect individuals to raise their hands if they have questions. A questioner is more likely to look expectantly at you—thus sending a more subtle message. Citizens of many countries consider pointing in public to be rude. Therefore, we should do some research before interacting with people from other nations [15].

If you are a college student or young practitioner, you are very proficient at using social media for faceless communication. As a person who spent the first half of his career in the pre-internet world, I value the effectiveness and efficiency of texting, emailing, blogging, and other uses of social media. However, like all technologies, social media has positive and negative features. English professor Mark Bauerlein warns us about "the avalanche of all-verbal communication," saying that it may diminish our communication ability. He argues that, because of excess use of faceless communication by some of us, we may be less likely to read body language, an essential part of interpersonal communication. Those lapses could harm us and our organizations [20].

The preceding body language discussion focuses on what you can learn from the body language of others. Briefly contemplate how your body language influences both others and you. For example, you, an engineering student, are about to meet with the CEO of a local manufacturing company. She agreed to let you interview her for a half hour so that you could complete a class assignment. You are a bit nervous but have done your homework about the company and the CEO and developed a list of questions. You have dressed appropriately, are well groomed, and wear shined shoes. After a short wait, an assistant escorts you to the CEO's office you walk in confidently.

Why confidently? Your company knowledge, your preparation, and your appearance tell you—as well as her—that you are ready to make the best use of your and her time. In his case, your body language consists of what you are wearing and your confident stride and posture. You established, partly through body language, a positive first impression.

2.5 ASKING AND LISTENING ADVICE

Let's further explore and connect asking and listening by offering some pragmatic advice for enhancing the effectiveness of the powerful asking–listening process. I remind you of the six principles of effective communication presented in Section 1.5. Keep them in mind as you consider the following asking and leading advice.

2.5.1 Prepare a List of Questions

Many of our interactions with others, whether F2F or virtual, are impromptu or routine, and most of these are enjoyable and fruitful. As you participate in these common events, with colleagues and others, ask questions, listen to the answers, and look for facts and feelings.

Other interactions, again, whether in-person or virtual, are scheduled and intended to address one or more topics. Some examples that you could encounter as a student or practitioner are:

- A meeting with your faculty advisor
- An interview with several faculty members for admission to a graduate program
- An interview with a potential employer
- A meeting of the project team of which you are a member
- A breakfast discussion with a client or customer

- A discussion with a consultant retained by your organization
- An introductory meeting with several executives and managers who work for a potential client or customer

Assuming you know the person or persons you will be communicating with, prepare a list of questions that you could draw on during the event. You are likely to be able to use some of them to help make the session mutually productive. If this will be your first contact with the person or persons you will meet within the scheduled event, do some research to learn what you can about them and their organization.

Informed by experience, I know that preparing a list of questions, and doing other research, as appropriate, adds to the success of all parties participating in a discussion, meeting, or interview. Recall the positive outcome of the situation I described, which involved 24 questions (Section 2.2).

Consider the following example of the effective use of a list of questions. An up-and-coming engineer, let's call him Joe, was promoted from project manager to department head in an engineering firm. Marketing engineering services was now one of Joe's major responsibilities, and things were not going well. The firm asked me, as their consultant, to assist Joe.

They scheduled two days for Joe and me to work together. We met early the first day, and I soon discovered that Joe was smart, competent, and enthusiastic about his work. That afternoon, I accompanied him to a meeting with a client, where he gave a project status report. Joe was an effective presenter. Why was this smart, competent, enthusiastic person, with presentation ability, not being successful in marketing services to new clients?

Late that first day, Joe mentioned that he would meet the next morning with a potential new client. Joe's firm had received a request for proposal (RFP), and Joe was in the process of preparing the proposal. An RFP is a document distributed by organizations seeking consultant assistance. It describes a potential project and asks interested consultants to indicate how they would approach the project, and often requests an estimated cost for services. Joe asked me if I would like to join him in the meeting with the possible new client, and I agreed.

We began to prepare for the meeting. Joe took out a pad of paper and said something like, "Let's write down everything we want to tell the client about our firm." That's when I "saw the light." I suggested, instead, that we write down questions along the theme of everything we would like to know about the potential client and the project. Joe somewhat reluctantly agreed, and we prepared the list with the idea that we would use at least some of the queries. I suggested this approach because the upcoming meeting should focus on our understanding of the potential client—facts, feeling, wants, needs—and not necessarily on them understanding our engineering firm.

The next day, as we drove to the potential client's office, I suggested that, by using our questions, we strive to have the client talk at least 75% of the time.

Joe agreed. We asked most of our questions during the one and a half-hour meeting, the two client representatives talked most of the time, we learned much, and Joe and his firm won the project.

Jeffrey Fox, in his book *How To Become a Rainmaker,* stresses preparing what he calls a pre-call plan for a sales call, which is a specific kind of asking and listening situation. According to Fox, "Ninety percent of all sales are won or lost before the salesperson sees the customer. This is because so few salespeople actually plan the call." For emphasis, Fox says that preparing for an important conversation is like a football coach preparing a game plan or a pilot performing a preflight check [21].

2.5.2 Avoid Dumb Questions

The statement, "there are no dumb questions," is sometimes offered as a way of encouraging a reluctant questioner to ask more or better questions. While the helpful intent may be admirable and appreciated, the statement is "dumb."

I say this because, as suggested in the previous section, when we are preparing to meet with new people in new situations, we should first find out what we can about them and their organization(s). Then, based partly on what we learn, prepare a list of questions to draw on as needed.

Failure to do the research can lead to question lists that include "dumb" queries such as:

- What does your firm manufacture?
- Where is your corporate office?
- What is your agency's function?

Questions like these indicate lack of preparation and may suggest individual or organizational incompetence. You can, as part of your preparation, readily answer questions like these via the internet and by asking friends and colleagues for help. Then meet and use that time to ask, learn about, and share facts and feelings not readily and widely known. That in-depth information often leads to mutually beneficial results and relationships.

2.5.3 Speak Their Language

As you ask questions and listen to the answers, note keywords used by the other person or persons and begin to employ those words as the conversation continues. Maybe not exclusively, but frequently enough to indicate that you are paying attention, trying to understand what others say, and enabling them to understand you.

Consider some examples:

- Does your industrial engineer friend design cars, automobiles, or motor vehicles?

- Does a potential client want a consultant to conduct a study, investigate an issue, or perform an analysis?
- Does the environmental group refer to the natural area along a stream as a floodplain, wetland, or green corridor?

When you speak their language for the sake of effective communication, you will also expand your vocabulary.

2.5.4 Talk to Strangers

Our parents probably cautioned us to be wary of strangers, as in "don't talk to strangers." Now, as engineering students or practitioners, we can, with reasonable caution, consider this advice, "Talk to strangers" [2].

We frequently encounter people we do not know at work, within our community, waiting for a flight, or at a conference. Regularly take the initiative to start a conversation with a stranger. Will it always, or most of the time, provide useful information, lead to a new business or professional relationship, or generate some other benefits? No, but as hockey great Wayne Gretsky said, "You miss 100 percent of the shots you never take."

I tend toward introversion, as do the majority of engineers. Nevertheless, I've made the effort to approach strangers in a variety of professional and other settings and, as a result, have enjoyed positive results. Some examples:

- At a conference, I introduced myself to a speaker after his presentation, complimented him on the content, and noted our apparent shared interest. This started a chain of events that quickly provided me with a multiyear consulting arrangement.
- Near the end of a session at another conference, I raised my hand, was called on, and asked a question. Immediately after the session, another session participant and I met, briefly discussed my question, and exchanged business cards. He was the president of an engineering firm. We occasionally communicated for about six months, and then the firm retained me as a consultant.
- While waiting outside of a hotel for a cab, I connected with an individual who was part of the senior staff at a national engineering society. We shared a cab ride to the airport and discussed some of the society's initiatives. That led to my appointment to a new, major committee, which enabled me to make contributions and introduced me to some exemplary engineers.

Imagine that you and some other engineers you work with attend meetings of the local chapter of an engineering society. Several of you just arrived at the chapter's evening dinner meeting. When you are about to select a table, resist the tendency to sit with your work colleagues. Instead, sit at a table that includes strangers.

You can easily talk to colleagues almost anytime, but the opportunity to meet those strangers may never happen again. When we stay within our usual group, as comfortable as that may be, we are likely to miss opportunities to connect and learn. Our employer, our clients/customers, and we may not be served as well as they could be. See Appendix C, Section C.2, for examples of some starter questions you could ask someone you just met at a conference or meeting.

As noted earlier, Appendix C provides example questions for other common situations, and, at the personal level, Section C.8 suggests questions to ask yourself if you want to grow personally and professionally. The last section in the appendix provides examples of important information learned by me and consulting firm colleagues by asking questions—information not readily available but openly acquired by simply asking.

2.5.5 Focus on Individuals Who Have the Power to Say Yes

Sometimes we communicate to achieve a very specific goal, one that requires someone's support or approval. In these situations, our success in applying the questioning–listening process depends on whom we communicate with. We should heed the advice of Eleanor Roosevelt, diplomat, activist, and First Lady, who said, "Never allow a person to tell you no who doesn't have the power to say yes."

Consider a hypothetical example that illustrates the meaning of that advice—the importance of who we deal with. Imagine that you are an employed professional engineer (PE), whose specialty is mechanical engineering and you recently learned about a company whose innovative products impress you. Your research reveals that these admirable products originate in the company's R&D department, led by a nationally known PE.

Being proactive, you contact the company's human resources (HR) department to see if they are listing openings for engineer positions in the R&D department. An HR staff member indicates that there are no such openings. You are discouraged but then recall Eleanor Roosevelt's counsel.

While the HR person may be able to say "no," that is, indicate no opening for you, if there were an opening, that person could advise you of it but not say "yes" to you, that is, have the authority to hire you. The R&D department head is the decision-maker, the "yes" person.

Therefore, you take a different tack. You find the name, credentials, and address of the department head. You write a carefully composed and accurate letter that indicates your admiration for the company's innovative products, states your interest in innovation, and mentions two innovation courses you took as part of an ongoing part-time master's degree program. You conclude by asking to meet F2F with the department head to introduce yourself further and learn more about R&D at the company.

Within a week, you receive an invitation from the department head for an interview, which occurs one week later. You and the department head talk for an hour—ask each other many questions—and you meet some other R&D personnel. Before you leave, the director indicates that he is beginning to define a new position; your KSA set matches and further defines that position; and he offers you the job, which you accept.

You may say nice story but things don't happen that way. Did for me. Early in my career, nearing the end of three years of work in academia, I became interested in watershed planning using digital hydrologic–hydraulic models, created a model for an engineering firm, and happened on an article that described how a regional planning commission in a Midwestern state was using a consulting firm to do modeling. This was decades ago, when such modeling was in its infancy. I contacted the commission's executive director, expressed interest in using modeling in watershed planning, and suggested that maybe I could be part of the commission's effort.

The director immediately replied and invited me to an F2F meeting. We asked each other questions, and the director indicated his strong desire to bring modeling in-house, upgrade it, and use it on all watershed projects. Because of my equally strong interest in modeling, effecting change, and the director's goal, I said I would like to lead the effort. They hired me. Our staff evolved as we built a comprehensive set of models, which, with continuous improvement, we applied to many watersheds. We shared our pioneering work by making presentations and publishing papers [22–24].

Consider my purpose in sharing Eleanor Roosevelt's advice, the hypothetical story, and my story. If you have a strong desire to do a particular type of engineering or work for a specific employer, then politely, professionally, and proactively pursue your goal. Identify and learn about decision makers and do what you can to ask them varied questions and listen carefully to their answers. As someone anonymously said, "Don't be afraid to go out on a limb. That's where the fruit is."

2.5.6 Verify Understanding

We ask questions and then listen attentively and with empathy so that we understand others—comprehend facts and feelings. Sometimes we need to verify what we think we heard and saw. Here are some ways to confirm your understanding [14]:

- Ask a question that begins with active verbs, such as explain or clarify, as in, "please explain what you mean by. . ." You may have to interrupt the speaker diplomatically in order to do this, which indicates that you are listening carefully.
- Paraphrase the other person's message as in, "please let me put in my words what you seem to be advocating."
- Draw or sketch your understanding of the process the speaker is describing. For example, turn over the restaurant placemat and say, "Please let me sketch my understanding of the process you are suggesting."
- Follow up, after the discussion, with a text, email, or telephone call and ask for clarification.

An effective way to show that you cognitively and emotionally understand the speaker is to use some of the speaker's facts, information, and views in subsequent communications. This does not necessarily mean you are in agreement with that person but does suggest that you heard and respect the message.

2.5.7 Trim Your Hedges

Consider some advice for responding to questions. We engineers tend to be honest and helpful in our interactions with others. However, when taken to extremes, that openness can lead to over-explanation, which causes another person or an audience to conclude that we lack competence, confidence, or commitment.

Responding to Questions

When answering questions, we need to trim our hedges, which means "the ability to speak, write, and answer questions in a positive manner, to present our views without excessive qualifications, so the information is presented at a level appropriate for our intended audience" [25]. Note the focus on the other person or the audience. For example, at a project team meeting, don't answer a question about the required size of pump by starting with, "If I did the calculations correctly . . ." Doing calculations correctly is your responsibility.

Contrast that situation, where a conditional answer is appropriate. You are meeting with professional peers one of them asks about the potential foundations for a series of towers on an electrical distribution system. You say, "Based on preliminary site data, in my opinion, we will not have difficulty designing foundations; however, I am concerned about potential high ground water levels at some tower sites." The data are incomplete, and you are offering your opinion to professionals who understand foundation complexity.

If you were at a public meeting and responding to a question about the tower stability, the previous answer would not be effective because the audience is unlikely to understand the context. An appropriate answer would be, "We are collecting site data which we will use to design adequate foundations."

Talking to Ourselves

Often, when answering a question or wanting to share a thought, we are also talking to ourselves and tempted to use hedges, caveats, and other conditions. "What we say and how we say it influences other's opinions of us, our work, and the organizations we represent" and influences our subsequent actions [25]. We risk creating low expectations for ourselves and those we serve.

Consider some examples of low-expectations statements—contain hedges—with each followed by more positive high-expectation replacement [25].

- I hope to get started on the assignment this weekend—I will get started on the assignment this weekend
- I will try to get the memorandum to you by Friday—I will get the memorandum to you by Friday

- I will have to find the data—I will be glad to find the data
- Excuse my messy office—welcome to my place
- I was lucky—I set and achieved goals

Just one or a few different words make a big difference in how others perceive us and how diligently we follow up on commitments. To paraphrase Confucius, the Chinese teacher of wisdom: "For one word a person is often deemed to be wise, and for one word a person is often deemed to be foolish. We should be very careful about what we say."

2.5.8 Avoid Hurtful Words

Recall that our definition of communication (Section 1.2.1) includes "conveying ideas, information, and feelings." Then consider this feelings observation by scientist Albert Einstein: "Although words exist for the most part for the transmission of ideas, there are some which produce such violent disturbance in our feelings that the role they play in the transmission of ideas is lost in the backgrounds." So true—words can hurt.

I was the tall, clumsy, and least effective player on my high school basketball team. During an early-in-the-season practice, one of the starting five players, frustrated by my ineptitude, said, "Walesh, you're tall and that's all." Those words hurt then, and even today, I feel the sting [26].

Whether we are asking a question, responding to one, writing a report, or speaking to an audience, let's be mindful that carelessly used words can hurt and, as a result, diminish or destroy our message and damage relationships. I am not advocating political correctness but am urging respectful and civil behavior.

Given human nature, you, as a student or practitioner, may receive or send a message that is hurtful to you or someone else. Try to resolve the issue with an F2F or similar conversation as opposed to using more impersonal social media. For additional thoughts about resolving interpersonal conflict, see Section 3.5.1.

2.5.9 Document and Share

As noted near the beginning of this chapter, the question-asking and listening process is powerful because both askers and listeners benefit; they learn from each other. If you follow the spirit and advice offered in this chapter, you will learn much about the people you interact with—facts and feelings.

Gradually develop the habit of documenting and, as appropriate, sharing what you learn. Assume, for example, that you, as a practicing engineer, are meeting with a potential client or customer in that person's office. As the two of you ask and answer questions, take skeleton notes.

As soon as possible after the conversation—while the exchange is fresh in your mind—review and expand your minimal notes. For example, if you drove to the meeting, expand the notes as soon as you get into your car—I frequently do this. Later, leverage what you learned by sharing your notes via email or a memorandum with others in your organization who may be interested, such as your department head or someone in marketing. If you habitually document important conversations and selectively share what you learned, others will reciprocate.

2.6 TIPS FOR SPECIFIC ASKING AND LISTENING SITUATIONS

Having just offered broadly applicable asking and listening advice, consider some practical tips for use in specific asking and listening situations.

2.6.1 Question and Answer Session After a Presentation

Most engineers will, probably beginning as students, be questioned after giving presentations in what are commonly called Q&A sessions. Other situations arise such as when several engineers question an engineer, say during a meeting. Because Q&A sessions are an integral part of speaking, Section 4.5.6 of the speaking chapter discusses them in detail. Refer to that section for advice on how to get the Q&A session started and how to respond to different kinds of queries.

2.6.2 Interviewing as Part of Applying for a Position

Assume that you, a senior engineering student, considered various postgraduate employment possibilities, did preliminary research about each, created a short list of potential employers, submitted inquiries or applications, and received an invitation to an on-site interview. That employer might be a business, government entity, educational institution, or other. Let's explore some ways for you to prepare for and participate in the interview.

Prepare a List of Questions

Review and expand your preliminary research about the potential employer. Learn more about the organization's history; vision, mission, and other value statements; strategic plan; services and/or products; size, such as number of employees; headquarters and office locations; and recent major accomplishments. Then create a list of questions to take to the interview, the answers to which would add to what you already know about the potential employer. Your list of questions might address topics such as:

- Challenges being faced
- Where you would work and who you would report to
- What functions you would fulfill, and what projects or other activities you would work on
- Roles of engineering licensure and various certifications
- Assistance with your professional growth

- Ways in which you might advance
- Compensation and benefits

Mix closed-ended and open-ended questions. Recall Section 2.2, which concludes that when you ask thoughtful questions during the interview, you will reveal your high-level preparation and strong interest in the organization.

Kinds of Questions to Expect

If you conduct a search for job interview questions, you will find many lists of questions typically asked by interviewers, with the number of questions ranging from 10 to 100. As an introduction to this topic, consider this short list, adapted from the *Harvard Business Review* [27]:

- Tell me about yourself and your background
- How did you learn about this position?
- What kind of work culture do you prefer?
- How do you handle pressure?
- Do you prefer teamwork or working independently?
- How do you organize working on multiple projects?
- How did you improve your capabilities during the past year?
- What is your expected level of compensation?
- Are you also pursuing other jobs?
- What was the reason for the gap year in your resume?

Given that you are probably applying for an engineer position, expect questions about your technical knowledge and skills and how you expect to use and expand them.

Hypothetical and Behavioral Questions

The interviewer, in trying to determine your values and how you approach challenges, may pose two fundamentally different kinds of queries—hypothetical and behavioral [27]. Consider an example of each:

- **Hypothetical:** "What would you do to encourage out-of-the-box thinking on your team?" and "How would you set and implement a personal goal?"
- **Behavioral:** "Give me an example of how you encouraged out-of-the-box thinking on your team" and "Provide an example of how you set and implemented a personal goal."

The behavioral approach elicits concrete, historic examples, which reveal much more about you as a job candidate than the hypothetical approach. Knowing what you did, that is, your behavior, provides more value than hearing what you say you would do. Realize that a candidate, in responding to a behavioral inquiry, could fabricate a response. However, an experienced interviewer could ask probing questions that would reveal the candidate's dishonesty.

Unless you know otherwise, be prepared for hypothetical and behavioral queries, giving most attention to the latter. That preparation will be insightful because you will revisit and learn more from the many challenges you have already faced in your personal, student, and professional lives.

Show and Tell

An engineering firm client asked me to help interview a short-listed candidate for a project manager position. After studying the candidate's resume, I interviewed him and asked many questions, seeking more information about positions held, projects he worked on, and software he used. I recommended hiring him. They did, and to our dismay, the new project manager soon demonstrated that he could not write legible letters, memoranda, and reports.

This embarrassing interview experience underlined the importance of a carefully designed interview process that begins with a thorough description of the KSA required for each position. In some cases, the best way to determine if a candidate has a certain specific KSA is to ask them to show or demonstrate—in show and tell manner—that they have them. For example [28]:

- If writing ability is crucial, ask a candidate to provide examples of letters, memoranda, and reports. Keep in mind their need to respect client confidentiality.
- If speaking is critical, require a candidate to make a presentation. As an applicant for an engineering dean position, I was asked to give a presentation to engineering faculty but present it as though I was speaking to students.
- If knowing how to approach a potential new project is part of a position description, then give the candidate a current RFP; ask the candidate to take an hour to skim it; and suggest some ways to respond.

Show and tell is not routine. However, I am sharing its purpose and means with you, just in case.

2.6.3 Taking an Oral Examination

Depending on your formal education and career path, a panel of experts may examine you at critical junctures. For example, near the end of an engineer's PhD program, a committee of faculty who are experts in the candidate's chosen specialty will examine the candidate orally. They will tend to focus on the candidate's largely complete research project and the draft written thesis that describes it.

Applications for some engineering positions may be subject to an oral examination conducted by a panel. When one or several members of an engineering firm present a project proposal to a potential client, a team of client representatives often questions them.

For the purpose of this discussion, I offer advice applicable to a PhD oral examination because that is the most demanding form. If you are preparing for some other kind of oral examination, you can get useful tips from the following suggestions based on my experiences and other sources [29, 30].

Before the Examination

1. Determine the structure of or agenda for the examination and the likely range of topics. Obtain that information by talking with students, faculty members, including some or all committee members, and your advisor.
2. Although you already know the committee members, search for more information about each of them using the previously mentioned sources. Learn more about their education and careers, expertise, favorite professional authors, personalities, and possible eccentricities. Based on those expanded profiles, imagine what may interest each of them most about you and your research.
3. Create a list of questions that you think committee members may ask. Here are some questions to get you started:

   What motivated you to select your research topic?

   What is the most original aspect of your research findings?

   What are your research strengths and weaknesses?

   If you were to start over, what would you do differently?

   Who should know about your research results?

   What graduate or undergraduate course or courses were most helpful to your research?

   If you could continue your research, what would you do next?

   What are your near-term career plans?

   How could we have been more helpful during your research?
4. Practice answering the questions aloud because of the benefits of out-loud practice, as described in Section 4.4.10.
5. View the scheduled and far-off examination as an opportunity to review your graduate courses and the motivation for and progress made on your research. As you do this, consider taking free hand, as opposed to typed, notes. Most studies of student note-taking conducted in the United States and other countries conclude that students who take notes by hand learn more and retain more than students who take notes with a computer [31–34].
6. Continue to read recent articles, papers, and books relevant to your research so your knowledge is current.

During the Examination

1. Take comfort in the thought that committee members want you to succeed—they invested in you as a student and researcher.
2. Dress appropriately—at least as formal as the most well-dressed member of the committee.
3. Make sure you understand each question before starting to answer it.
4. Don't over answer questions—focus on what you think is the asker's interest.
5. Take skeleton notes, in which you record apparent committee member concerns and points you want to make.
6. Disagree politely if you think a committee member errs—you and they are collectively seeking the truth.

2.7 KEY POINTS

- Proactive asking and listening help us balance our natural desire to be understood by others with our equal, if not more important need, to understand them.
- Asking and listening provide benefits such as an obligation to respond; broader and deeper thinking by participants; sharing of data, information, and knowledge; temporarily putting the asker in the driver's seat; and self-persuasion.
- The most common question-asking obstacle appears to be fear of appearing uninformed or poorly prepared. Other obstacles include reluctance to question authority, coming across as intrusive or rude, and not knowing what or how to ask.
- The chapter describes five very different question-asking methods.
- To understand others fully—to achieve empathetic listening—we should go way beyond simply hearing and strive to listen and look, the latter via body language, for facts and feelings.
- The chapter concludes with broad and in-depth pragmatic advice for initiating and conducting the powerful asking–listening process and offers some tips for specific and common asking and listening situations.

Nature, which gave us two eyes to see and
two ears to hear,
has given us but one tongue to speak.

—*Jonathan Swift, Irish satirist*

REFERENCES

1. Leeds, D. (2000). *The 7 Powers of Questions: Secrets to Successful Communication in Life and Work*. New York: Berkley Publishing.
2. Walesh, S.G. (2012). *Engineering Your Future: The Professional Practice of Engineering*. Chapter 14, Marketing: a mutually-beneficial process (p. 421). Hoboken, NJ: Wiley.
3. Lyons, D. (2023). Mute: overtalkers are everywhere—but saying less will get you more. *Time*. January 30/February 6.
4. Walesh, S.G. (2018). How to ask the right questions to achieve win-win results. ASCE. Webinar presented 14 January 2018.
5. Walesh, S.G. (2014). The five habits of highly effective marketers. ASCE. Webinar presented 3 September 2014.
6. Covey, S.R. (1990). *The Seven Habits of Highly Effective People*. Habit 2: Begin with the end in mind. New York: Simon & Shuster.
7. Kipling Society. (2023). I keep six honest serving men. https://www.kiplingsociety.co.uk/readers-guide/rg_servingmen1.htm (accessed 6 January 2023).
8. Liker, J.K. (2004). *The Toyota Way: 14 Management Principles from the World's Greatest Manufacturer*. (pp. 252–254). New York: McGraw-Hill.
9. Walesh, S.G. (2017). *Introduction to Creativity and Innovation for Engineers*. Chapter 4, Basic whole-brain methods. (p. 139). Hoboken, NJ: Pearson Education.
10. Sittenfeld, C. (2000). The most creative man in Silicon Valley. *Fast Company* (June).
11. Wikipedia. (2023). Echo question. https://en.wikipedia.org/wiki/Echo_question (accessed 7 January 2023).
12. Merriam-Webster. (2023). Dictionary. https://www.merriam-webster.com/dictionary/hearing (accessed 8 January 2023.
13. Linver, S. (1978). *Speak Easy*. New York: Summit Books.
14. Walesh, S.G. (2012). *Engineering Your Future: The Professional Practice of Engineering*. Chapter 3, Communicating to make things happen (pp. 77–78). Hoboken, NJ: Wiley.
15. Herman, A.E. (2016). *Visual Intelligence: Sharpen Your Perception, Change Your Life* (p. 133). Boston: Houghton Mifflin Harcourt.
16. Hall, E.T. (1990). *The Silent Language*. New York: Anchor Books.
17. Emerson, R.W. (1876). The conduct of life. https://archive.vcu.edu/english/engweb/transcendentalism/authors/emerson/essays/behavior.html (accessed 11 January 2023).
18. Quilliam, S. (2009). *Body Language: Actions Speak Louder Than Words*. New York: Fall River Press.
19. Wang, J. (2009). Non-verbal cues and what they (probably) mean. *Entrepreneur* (May).
20. Bauerlein, M. (2008). *The Dumbest Generation: How the Digital Age Stupefies Young Americans and Jeopardizes Our Future*. New York: Jeremy P. Tarcher/Penguin.
21. Fox, J.J. (2000). *How to Become a Rainmaker: The People Who Get and Keep Customers* (pp. 14–17). New York: Hyperion.
22. Walesh, S.G. (1973). Simulation in watershed planning. *Journal of the Hydraulics Division*. ASCE. September.
23. Walesh, S.G. and Videkovich, R.M. (1978). Urbanization: hydrologic-hydraulic-damage effects. *Journal of the Hydraulics Division*. ASCE. February.
24. Walesh, S.G. and Raasch, G.E. (1978). Calibration: key to credibility in modeling. *Proceedings, Hydraulics Division Conference, Verification of Mathematical and Physical Models in Hydraulic Engineering*. ASCE. August 1978.

25. Walesh, S.G. (2004). *Managing and Leading: 52 Lessons Learned for Engineers.* Lesson 19, Trimming our hedges. Reston, VA: ASCE Press.
26. Walesh, S.G. (2004). *Managing and Leading: 52 Lessons Learned for Engineers.* Lesson 17, You're tall and that's all. Reston, VA: ASCE Press.
27. Oliver, V. (2012). 10 common job interview questions and how to answer them. *Harvard Business Review.* https://hbr.org/2021/11/10-common-job-interview-questions-and-how-to-answer-them (accessed 30 June 2023).
28. Walesh, S.G. (2004). *Managing and Leading: 52 Lessons Learned for Engineers.* Lesson 46: Interviewing so who you get is who you want. Reston, VA: ASCE Press.
29. Hallstein, E., Kiparsky, M., and Short, A. (2023). An orals survival kit. Office of Graduate Studies. University of Nebraska. https://graduate.unl.edu/connections/orals-survival-kit (accessed 30 June 2023).
30. Higginbotham, D. Editor. (2022). 5 tips for passing your PhD viva. Prospects. https://www.prospects.ac.uk/postgraduate-study/phd-study/5-tips-for-passing-your-phd-viva (accessed 1 August 2023).
31. Anadale, C. (n.d.). Why you should take notes by hand. YouTube video. https://www.youtube.com/watch?v=eTxZTZskVFQ (accessed 1 July 2023).
32. Bothwell, E. (2017). Pen and paper beats computers for retaining knowledge. *London Times* Education Supplement, February 13. https://www.timeshighereducation.com/news/pen-and-paper-beats-computers-retaining-knowledge (accessed 1 July 2023).
33. Doubek, J. (2016). Attention, students: put your laptops way. *Scientific American.* April 17. https://www.npr.org/2016/04/17/474525392/attention-students-put-your-laptops-away (accessed 1 July 2023).
34. May, C. (2014). A learning secret: don't take notes with a laptop. *Scientific American.* June 3. https://www.scientificamerican.com/article/a-learning-secret-don-t-take-notes-with-a-laptop/ (accessed 1 July 2023).

EXERCISES

2.1 Researching a potential employer: Select an organization that is high on your list of potential employers. It could be a manufacturer, government entity, consulting firm, educational institution, engineering society, or whatever fits your situation. Given that you favor this employer, you already have some information about it. Now dig deep and wide so that you learn much more. To get you started, here are some suggested topics: history, mission, vision, strategic plan, size, locations, and services and/or products. These are just ideas for your consideration. Document what you learn.

2.2 Questions for an interview: Given what you have learned about the potential employer in Exercise 2.1 and, assuming an invitation to an initial F2F or virtual interview, prepare a list of questions you want to ask. Include closed-ended and open-ended questions. For some question ideas, see Section 2.6.2.

2.3 Interpreting body language: As you go about your activities the next few days, look for opportunities to observe the body language of speakers. Examples of speakers: student making an in-class presentation, instructor teaching a class, colleague making a presentation at a meeting, and individual expressing views to friends. Using a two-column format, list in the first column an observed example of body language and, in the second column, the positive or negative impact on you. No need to name the speakers.

I offer this exercise for two purposes. First, illustrate explicitly how listeners and viewers are influenced, often unknowingly, by body language. Second, to suggest how we, as speakers, often unknowingly influence others via our body language. Privately reflect on the positive and negative signals you send and how you can capitalize on the former and diminish the latter.

2.4 Evaluating, asking, and listening advice: Review Section 2.5 and determine what is the most valuable advice for you right now. Write a brief summary of what you selected and why.

2.5 Interview an excellent communicator: Draw on your knowledge, ask around, and identify one or more excellent communicators—within or outside of engineering—on your campus, in your company or agency, or within your community. Select one person, contact that individual, and ask if you could have a half hour of their time for an F2F interview. Consider explaining that the interview is a learning exercise for the course you are taking. As you seek an interviewee, prepare some questions, including closed-ended and open-ended ones. Conduct the interview and prepare a written summary with an emphasis on what you learned about communication and other matters.

CHAPTER 3

WRITING

Hard writing makes for easier reading.

—Thomas J. Brown, management writer and speaker

After studying this chapter, you will be able to:

- Explain why writing is an essential part of engineering
- Describe the fundamental differences between writing and speaking
- Provide examples of two communication situations, one where writing is likely to be more effective and one where speaking would be the best choice
- Discuss how you can use writing to learn
- Select one or more of the many writing advices offered in this chapter and show a writer who is less knowledgeable than you how to use them

3.1 WRITING IS AN ESSENTIAL PART OF ENGINEERING WORK

Recall this definition of communication presented in Chapter 1: the act or process of effectively conveying, from one person to one or more others, information, ideas, and feelings using asking, listening, writing, speaking, visuals, and mathematics. We addressed asking and listening in Chapter 2, now let's delve into writing.

As I started writing this chapter, with the goal of helping readers write well, I thought about the intended audience—mostly engineering students and, secondarily, practicing engineers. What documents and other items would they eventually write?

Experience indicates that you and other readers will, beginning as students, compose different kinds of documents, which will expand in number, variety, and importance as you enter and move up in engineering practice. Eventually, as practitioners, you are likely to compose frequently most of the following kinds of documents: texts, emails, memoranda, meeting minutes, letters, reports, specifications, manuals, proposals, resumes or curriculum vitae (CVs), articles, peer-reviewed papers, and books. Engineers often communicate, including writing, while doing technical tasks and then communicate, again often in writing, the results of those tasks.

The writing ideas, principles, advice, and tips offered in this chapter apply to many kinds of writing. I wrote this chapter with the hope that you will use some of its content on every one of your student and practitioner days because I know that you will write something every day. Frankly, given the importance of writing in your daily activities, this chapter can prove to be a powerful resource for you.

3.2 DIFFERENCES BETWEEN WRITING AND SPEAKING

A speaking chapter immediately follows this writing chapter. Therefore, this is the opportune time to describe fundamental differences between these two communication modes [1, 2]. By knowing these distinctions, you will know how to select the best mode for any communication situation and how to make optimum use of the chosen mode.

3.2.1 Writing: Single Channel and One Direction

As illustrated in Figure 3.1, writing is mostly single channel, that is, verbal—one word after another—and one-directional. I say "mostly single channel" because, although the writer and reader do not see each other, the writer can send some

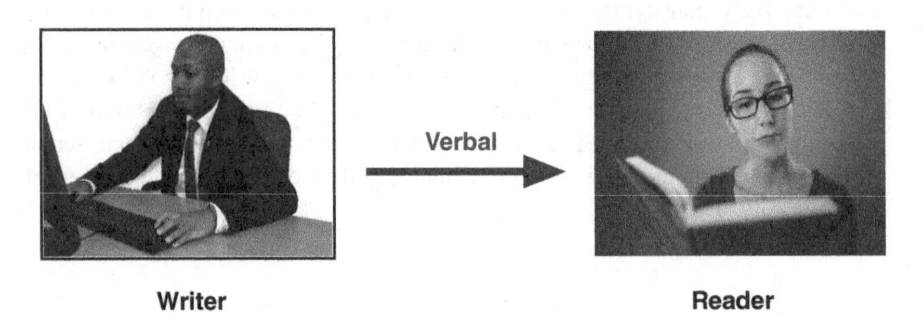

Figure 3.1 Writing is single-channel and one-directional. *Source:* S. G. Walesh and Pixabay.

modest visual content to supplement the words. Examples include tables, figures, and an attractive composition. However, the writer conveys the message mostly via carefully chosen words.

While the reader could reply, that is not an obligation. You or I can compose and send what we view as a thoughtful insight, excellent idea, or great solution and never hear from the recipient.

3.2.2 Speaking: Three Channels and Two Directions

Figure 3.2 shows how speaking, in sharp contrast with writing, uses these three channels:

- **Verbal**: words used by the speaker, just like with writing
- **Vocal**: how the speaker uses his or her voice as defined by speed and intonation, that is, how fast the speaker talks and the variation in pitch, that is, frequency of the sound
- **Visual**: body language, visual aids, and props

Speaking also guarantees some instant feedback from the audience because speaking is two-directional, as also shown in Figure 3.2. The speaker can judge the receptiveness of an audience member by what they say (verbal), how they say it (vocal), and their facial expressions or other body language (visual).

3.2.3 Implications of the Differences Between Writing and Speaking

Being essentially single channel and one-way means that writing is a much more difficult means of communication, especially if you are trying to exert influence. Why? Only one channel to send the message and no expectation of immediate, up to three-channel feedback. Therefore, when writing, we need to get it right the first time and not have unrealistic expectations.

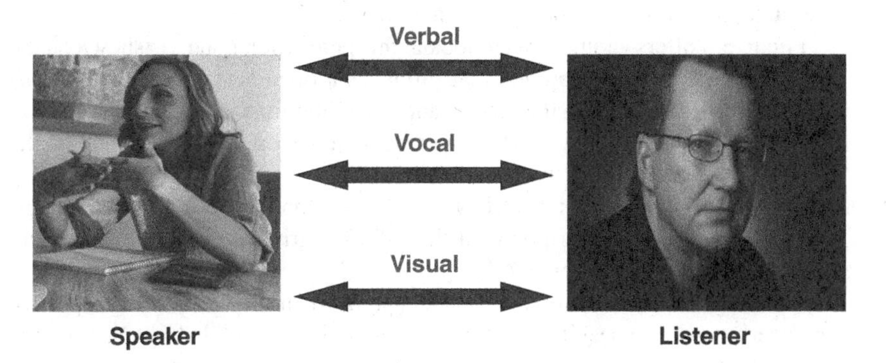

Figure 3.2 Speaking uses three channels and is two-directional. *Source:* Used with permission of ASCE. [2], images by Pixabay.

Yet, given the choice, I suspect many of us would rather write a message to a large number of people than present it orally to those people assembled in an audience. I say this because of widespread discomfort with public speaking, a topic discussed in the next chapter.

Generally, if our primary communication goal is to transmit data or information to one or more individuals, then writing is the preferred mode. In contrast, if we are advocating major change or a completely new creative or innovative approach, then use speaking at the outset. Speaking is more persuasive because it offers three channels and is two-directional. For example, assume you want to implement a completely new way of doing something in your organization. First, speak one-on-one, with individuals that you think would be interested in your idea and have the authority to implement it. If those conversations confirm interest and possible support, follow up with a written description of your proposal and request assistance [1, 2].

Consider an example of when I, as a middle manager in an engineering firm, got it wrong. I wrote a detailed proposal suggesting one-month sabbaticals for senior professionals, subject to certain criteria, and sent it to company's president. No response. In retrospect, I might have been successful if I had first met with the president, explained the idea, assessed his level of receptivity, and then followed up with a written proposal that, among other things, addressed his concerns.

3.2.4 Whether Writing or Speaking, Do it Well

Whatever communication modes we select, strive to use them very well. We don't want our team to carefully collect and analyze data in an "A" manner and then share the results in a written report that earns, in the minds of readers, a grade of "C." Those readers are also likely to give the data collection and analysis a "C," thus largely negating the team's efforts. Same thinking applies to a great "A" idea conceived within our organization and then presented via a "C" speech to an assembled group of decision makers [2].

Figure 3.3 offers another way to look at this perception issue. As shown on the left, when working on a project that produces a document, we invest the vast majority of our time doing the project and very little time documenting what we did. However, as indicated on the right, readers judge our effort based not on the work but on the quality of the documentation.

As noted by author Michael Lewis, "People don't choose between things, they choose between descriptions of things" [3]. Strive to produce an excellent description of your excellent work.

As I argued at the outset of this chapter, writing is an essential part of engineering—so is speaking. This and the next chapter will help you earn good to excellent "grades" in writing and speaking to go along with and help to implement your "A" technical work.

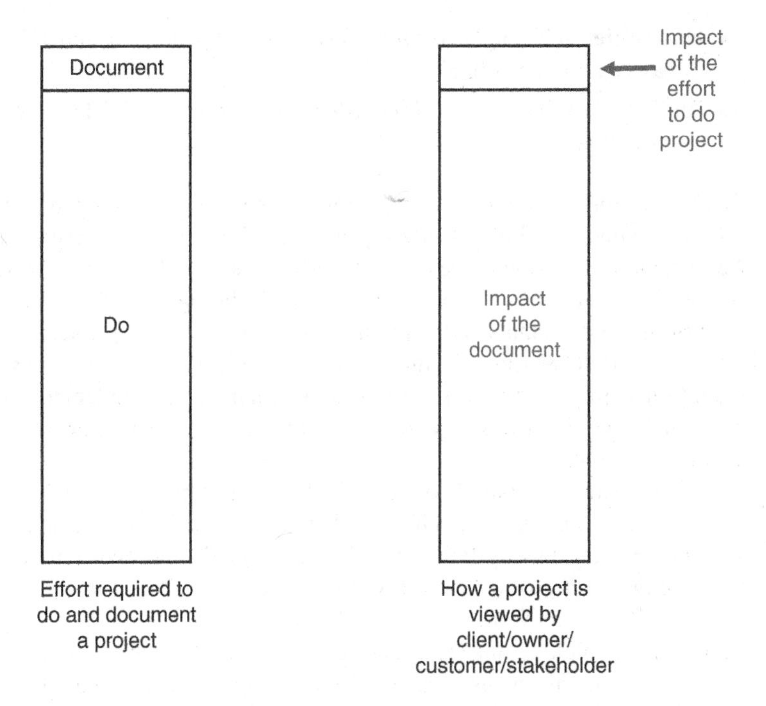

Figure 3.3 The document describing your work will have much more impact on those you serve than your effort in doing the work.

3.3 LEARNING TO WRITE OR WRITING TO LEARN?

My wife and I enjoy browsing at garage sales. Upon arriving, we separate, her to various treasures and me to the inevitable shelves, piles, or boxes of books. I recall looking into a box and seeing—or thinking I was seeing—the book *Learning to Write*. I reached for it and was surprised to see the title, *Writing to Learn* [4]. As suggested by the title, the author argues that students learn more about any topic when they write about it. Refer to Section 1.3.3 to review how writing provides opportunities for learning what we know, don't know, and need to know and for self-reflection.

For further insight into the reflection and learning value of writing, consider these thoughts by various writers:

- Edward Albee, playwright—"I write to find out what I'm thinking"
- Stephen King, author—"Writing is refined thinking"
- Dylan Thomas, Welsh poet—"The blank page on which I read my mind"
- William Zinsser, writer, editor, teacher—"Writing is thinking on paper"

- Eric Hoffer, self-taught social philosopher—"You learn as much by writing as you do by reading"
- Ralph Waldo Emerson, schoolmaster, minister, writer—"The ancestor of every action is a thought"

The title of this section poses a question. I suspect you would agree that the answer to that query is "both." Certainly, we should see and value writing as one of the five communication modes we are studying, especially in this chapter.

In addition, and this is the point, as you work through the chapter learning about writing fundamentals and applying writing advice and tips, keep in mind the potential parallel process of using writing to learn. Whether you are a student or practitioner, proactively seek writing assignments as a supplement to your formal education. Find ways to write about and, therefore, learn more about topics that interest you.

Consider a personal example of proactively using writing to learn. Years ago, I was increasingly thinking about how civil engineers' work would change as we move well into the twenty-first century. Who will do the civil engineering? Whom will those future engineers serve? What needs will they fulfill? How will the way civil engineers work change?

To focus my thinking and learn answers to those questions, I committed to writing a paper on this change topic and presenting it at a national conference. The paper and its presentation would be part of a conference session devoted to looking into the future. The conference organizers accepted the proposal. I quickly and happily accelerated my research and moved rapidly up the learning curve. I presented the paper and subsequently applied my new knowledge in various ways, including as a major section in one of my books. My story is an example of how to "get on the program," which is discussed in detail later (Section 4.4.1).

3.4 WRITING ADVICE

We turn now to practical advice for improving your writing. This is the perfect time to review Section 1.5, "Principles of Effective Communication." The first three principles—know audience, state purpose, and accommodate preferred ways of understanding—are essential and assumed in this large advice section.

View the following writing advice as resources to help you write more effectively. Don't expect to use all of them for any given writing project. View the advice as a smorgasbord, and choose and apply those that seem most appropriate. Use the suggestions I offer to plan your next major writing endeavor, that is, plan your work and then work on your plan. Don't start writing before you know where you are going.

3.4.1 Apply Existing Style Guide or Prepare and Use One

Recognize that many documents have multiple authors, each of which brings writing preferences. For example, I worked on a project for the Indiana Department of Natural Resources (IDNR). One of my responsibilities was to facilitate the preparation of a report by a team.

From the outset, I saw inconsistencies as various writers weighed in. For example, the sponsoring organization was, depending on who was writing, the Indiana Department of Natural Resources, the Department of Natural Resources, the Department, the IDNR, or DNR. Accordingly, I led the preparation of a short project-specific style guide for this project. Excerpts, which appear in Appendix D, suggest the simplicity and usefulness of the project-specific guide.

A style guide helps achieve consistency within a team-written document. It also encourages consistency among documents written within a single organization and even within a single document written by one person. A style guide's overall purpose is consistency and clarity, which aids readers by minimizing unnecessary distractions.

Style guides differ widely in content and length. Some content examples: abbreviations, capitalization, citations, compound words, headings, and subheadings, often-confused words and expressions, often-misspelled words, punctuation, quotes, use of numbers, and word processor settings.

If you, as a student or practitioner, receive a major writing assignment, ask if the organization has an applicable style guide. When I started writing this book, the publisher Wiley provided their style guide. Some engineering firms, government entities, and universities develop and use style guides.

If an organizational guide is not available, consider preparing a short, project-specific guide—once in use, its value might prompt the gradual development of an organization-wide guide. You could use the current version of the widely recognized *The Elements of Style,* originally published in 1919 [5]. While not tailored to a particular discipline, profession, or type of organization, you could supplement this helpful guide with project-specific or organization-specific provisions. The massive (1150 pages), comprehensive, and detailed *Chicago Manual of Style* will meet the needs of organizations that produce numerous and varied publications [6]. Finally, Colorado State University's (CSU's) open-access writing guides, although not intended to be just style guides, could provide input to your style guide [7].

3.4.2 Generate Content Ideas to Get Started

Even though I have written and published many essays, articles, peer-reviewed papers, manuals, reports, and books, I still feel apprehensive when beginning a new writing project. Whether I start with pen and paper or a word processor, the blank page or screen intimidates me. I may be tempted to procrastinate. Writer Peter De Vries was probably including this getting-started anxiety when he said, "I love being a writer. What I can't stand is the paperwork."

You too may feel anxious and be tempted to procrastinate when asked or given the opportunity to write something substantive—a document more complex than, say, an email or short memorandum. Therefore, I offer three processes to help you get started. They differ in detail, and the first two draw explicitly on neuroscience. All three processes will produce an initial outline of the document.

Method 1—Brainstorm, Incubate, Cluster, and Outline

Place a blank sheet of paper on your desk or bring up the word processor on your computer. Quickly think about ideas and information potentially connected to the document's purpose and audience. Capture in a single or few words, placed anywhere on the paper or screen, almost anything that "pops" into your mind. Don't over evaluate your thoughts. Simply record them and avoid writing sentences. One idea or piece of information tends to inspire another one. Brainstorm for no more than 15 minutes and look at the result—your initial "brain dump." While it may appear chaotic, it contains the basic building blocks of your intended document.

1. Take a break—maybe a walk around the block, if, for whatever reason, you cannot get going with the preceding brainstorming. Then resume brainstorming. The German novelist Thomas Mann said, "Thoughts come clearly while one walks." William Wordsworth, the English poet, composed most of his poetry while walking [8].
2. Set aside the brainstorming results—do something else for an hour or even a day. This temporary and complete disconnect from your writing project is essential. You used your conscious mind to perform the initial "brain dump," and now you should allow your subconscious mind to broaden and deepen the results.
3. Consider the following very brief comparison of conscious and subconscious minds [9, 10]. I offer this introduction as part of describing the first method to emphasize that the first two getting-started methods presented here reflect current neuroscience.
   - We know when we are using our conscious mind, but we rarely know what our subconscious mind is doing.
   - We can turn off our conscious mind, as when sleeping, while our subconscious mind operates 24/7.
   - Our conscious mind can think of only one topic at a time, in contrast with our subconscious mind, which functions as a massive parallel processor—our subconscious minds do most of our thinking.
4. Revisit your original "brain dump," and as you scan it, consciously note new ideas and information that appear in your mind. Your subconscious mind, which has been working, unbeknownst to you, on your writing

project, delivers those ideas and the information. Add new ideas and information.

5. Set the entire enlarged "brain dump" aside at least one more time, then revisit it and expect to see even more content.
6. Begin to bring some order to your display of content by arranging the items into preliminary groups.
7. Use the groups to create a first and rough outline of your document. Resist being prematurely bound by the outline. It's just words on paper—freely delete, insert, and rearrange.

You have a long way to go because you have yet to write a single sentence. However, you are off to a good start, especially if you consider how little time you invested relative to the large and varied potential content you generated. Another thought is that the above process could also be used by a small cognitively diverse group, in which case even more potential content would be generated.

Method 2—Mind Mapping

This method, like the first one, begins with brainstorming. Method 2 differs from Method 1 in that Method 2 is more visual, which is advantageous because, as we will discover in Chapter 5, vision is our most powerful sense.

I will illustrate mind mapping by sharing an example from my work [11]. A few years ago, I committed to preparing text and visuals for a presentation about cold calls. Within the engineering consulting business, a cold call means that a representative of an engineering firm meets, for the first time, with a representative of a potential client. The former's long-term goal is to provide services to the potential client. The short-term goal is to begin earning the potential client's trust and then learn their wants and needs.

While I had made cold calls, I had never written or made a presentation about them. Therefore, I applied mind mapping, which I had used many times, to generate ideas and information for possible use. Figure 3.4 illustrates the resulting mind map.

I started by writing "Cold calls" presentation in the center of a piece of paper and then added whatever "popped" into my mind and drew an arrow from the initial note to the new item. Cold calls are notoriously unpopular, which led to "Ugh" as my first thought. As expected, like in the case of Method 1, one idea or bit of information motivated another one, which I added to the evolving mind map along with an arrow showing its origin.

I worked, in short spurts, for four days on the ever-enlarging mind map, investing less than an hour in the total effort. Once I had consciously constructed an initial mind map, I turned it over to my subconscious mind several times, each time revisiting and further expanding it. Figure 3.4 is the resulting mind map,

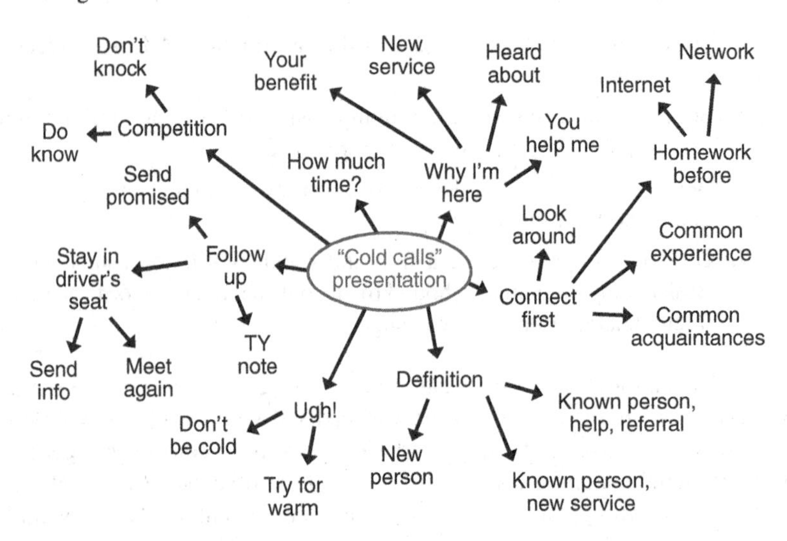

Figure 3.4 This mind map generated content ideas for a cold call presentation.
Source: Used with permission of Pearson [11].

with one exception. It shows all the text that I generated and all the interconnecting arrows. However, my original was not that neat.

You as a writer can, individually or as part of a cognitively diverse group, construct a mind map with very little effort and get a great return on your investment—content ideas for your document. Mind maps grow rapidly because they are visual and expand radially, not linearly. The existing words stimulate adding new words for new ideas.

As is the case with the first method, you can use a mind map to organize information into groups and use the groups to create a first outline of the document. Grouping is easier when starting with a mind map because the visual process used to create the mind map tends to result in formation of groups. For example, in the case of the cold call mind map, the item "Follow up" generated a group of related items. To conclude the example, I used the outline to prepare the text and visuals for the cold call presentation.

As a side note, I encourage you to explore individual and group use of mind maps to generate content for documents and presentations, explore possible causes of engineering failures, and identify potential solutions to engineering problems.

In summary, the first two get-started methods are similar in that each uses intentional conscious-subconscious mind interaction and produces an initial outline for a document. The outline is an invitation to start writing sentences.

A participant in one of my webinars shared a thought about use of his conscious and subconscious minds. He wrote, "I long ago developed the habit of rough drafting particularly sensitive letters or memoranda in the late evening;

then when I get back to them first thing the next morning, my subconscious often will have done a pretty good job of overnight editing."

Method 3—Artificial Intelligence

By coincidence, my researching for and writing of this book paralleled the rapid rise of interest in artificial intelligence (AI). ChatGPT, which stands for Chat Generative Pre-trained Transformer and is one small part of AI, launched in November 2022 and received major attention.

I experimented with ChatGPT and found that, when asked for content about a topic, it is a potentially useful content generator. Of course, I typically do not know all the source(s) used, the accuracy of what I receive, or the thoroughness of the response.

On reflection, AI places us in another period of advancing technology, during which we envision opportunities for a better quality of life, recognize the challenges, and eventually work things out. Current conversations about the potential positive and negative aspects of AI remind me of the 1970s discussions within engineering education about whether or not to allow use of the then new and costly hand-held electronic calculators—and, if so, how to use them.

However, given human nature and, as is usually the case with a new technology, some AI users will see it as an opportunity for unethical, illegal, and nefarious activities. Others will view AI as opportunity to increase productivity, creativity, and the quality of life.

The current interest in AI offers productive teaching-learning opportunities. Faculty and students can track the continuing development of AI and critically explore its likely positive and negative consequences as a writing tool. Those discussions could include the potential negative impact of unethical use of AI on the academic integrity of individual students and entire departments or institutions.

In the spirit of exploring new technology and helping you when writing, I suggest that ChatGPT, or similar AI systems, when used as a content generator, can help alleviate some of the "get started" stress faced by many engineering students and engineering practitioners when given a writing or speaking assignment. I offer this use of AI as Method 3, subject to the following suggestions:

- Frequently remind yourself that you are using AI to stimulate thinking about content— not to find and use content without confirmation
- Depending on the document and its intended readers, consider indicating that you used AI as an idea-generating tool
- When querying AI about a topic, ask it to provide sources
- If AI leads you to specific content that you want to use, find and cite the sources to verify the accuracy of the data, information, and ideas AI provides

- Avoid committing plagiarism—using the ideas or writings of others without crediting the source
- Determine the date of the last update of the AI system you use
- Recognize that AI has no conscience, morality, or ethics and maybe lacks the ability to be creative/innovative
- Appreciate that, the quality of content aside, AI has teaching and learning value in that it tends to produce grammatically correct sentences and coherent paragraphs and favors writing in the desirable active voice, as discussed later in this chapter
- Also, appreciate that, like most new technologies, AI is likely to evolve into an increasingly useful and legitimate tool for writing and other engineering endeavors

If you are an engineering student, expect some of your instructors to put even more emphasis on thoroughly researching a topic and requiring careful citing of sources. This approach will encourage using AI while discouraging explicit use of content that it produces, some of which may lack cited sources and be inaccurate and incomplete.

Going forward, I think that the engineering community will slowly and cautiously explore and use AI. Engineer employers in the business, government, and academic sectors and engineering societies will gradually develop policies and guidelines for AI use. Hopefully, consistent with engineering ethics codes, the highest priority in those policies will be public protection, as discussed in Section 1.1.4.

You could use AI to generate some content ideas for the substantive document you are writing. Query AI about the topic, sift and winnow the results, and use what remains to create the initial outline for your document.

***

Consider some additional thoughts about a document outline that you could draft using the results of any of the three just-discussed content-generation methods. Someone may advise you to skip using any of the three methods. Instead, start your document-writing effort by drafting the outline. While an outline is valuable because it begins to define the content and flow of the document, it is not, in my view, the place to start. Instead, begin the document-writing effort with one of the three what we might call brainstorming methods or something like them.

If you, at the outset of the document project, draft the content, it will reflect what you know about the topic and how you typically write reports. You will prematurely handcuff your writing endeavor. In contrast, if you first use, individually or as part of a small group, one of the three methods, you will generate more diverse and creative content ideas. Therefore, your outline, and the resulting document, will have more value.

3.4.3 Begin Writing Major Documents on Day 1

As engineering students and later as practitioners, we increasingly find ourselves responsible for leading the preparation of major documents. It may be a report about our teams' capstone course or senior project, our master's degree or PhD thesis, or our consulting firm's study of ways to improve a manufacturing process. The resulting document will be large, complex, have input from varied individuals and sources, and be of interest to many and varied stakeholders.

My advice: start writing that kind of document on day 1. Do I really mean on the first day of the project? No, not exactly, but I do mean during the first week of a three-month project and the first month of a year or longer project.

Let's first consider how to immediately start work on the document, that is, how to write the document in parallel with doing the work. After explaining the "how," we will review the resulting benefits [12, 13].

How to Start on Day 1

Assume you are the leader of a team that will work on a three-month project culminating in a report. The project will begin tomorrow. You, the team, primary beneficiaries, and stakeholders—individuals who will be, or think they will be, affected by the project's results—seem to understand the problem and the planned process to solve it.

During the first week of the project, you draft an outline or table of contents showing your current vision of how the report could be structured. You share it with team members and solicit their input, some of which may be very significant, as in "I didn't know we would be considering robotic devices." You markedly modify the outline, feeling confident that you and your team are now closer, literally and figuratively, to being on the same page. The lower part of Figure 3.5 illustrates this practice of working on the project in parallel with writing about it.

During the second week of the project, you send the current table of contents to selected stakeholders for review and comment. If the project were part of a capstone course, you would share the outline with your course instructor(s) and with external sponsors of your project. If the project were a search for ways to improve a manufacturing process, you, on behalf of your consulting firm, would send the table of contents to your client and, with client approval, selected stakeholders and request input. Expect, welcome, and reflect input—you want to discover misunderstandings, doubts, and new approaches now, when they can be efficiently addressed, not later, when costly remedial efforts would be required. You are continuing, as shown in the lower half of Figure 3.5, to both work on the project and write about it.

During the third week of the three-month project, you send a draft of the report's first chapter to your team. You may wonder how, so early in the project, you could write a chapter. Typically, the first chapter of a major document states

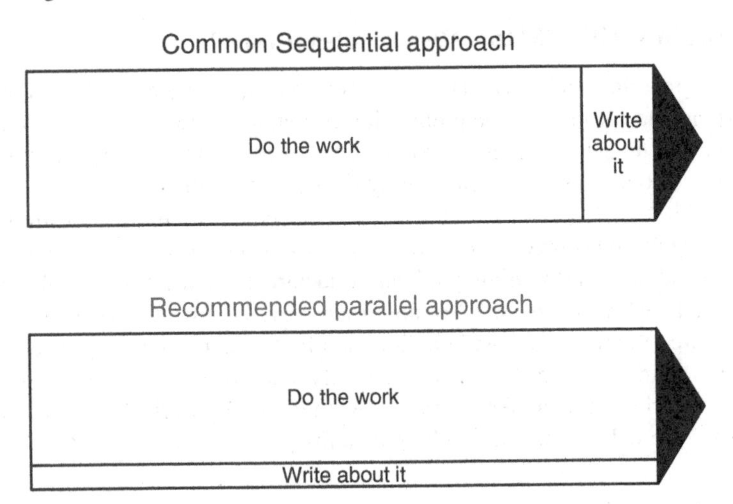

Figure 3.5 The recommended parallel approach to documentation reduces writing errors and enhances internal and external communication.

the project's purpose; provides background; defines key terms; describes the problem to be solved, explains the issue to be addressed, or opportunity to be pursued; and outlines the team's approach. If you cannot draft a chapter about those topics three weeks into the three-month project, your team may be spinning its wheels and wasting resources.

Continue the pattern of drafting portions of the document and sharing each portion, first with your team and then with stakeholders. I know from experience, first as a student and then as a practitioner, that engineers and others often prepare major reports using the sequential approach shown in the top half of Figure 3.5, that is, complete the project and then write the report. In stark contrast, and for your sake, I advocate the illustrated parallel approach.

***

In a related matter, a participant in one of my webinars anonymously shared his method for getting started on a report. He said: "One of the aides that I have prepared that helps me get a start on brief design reports is a Word template with just the title page and section headings (i.e., Purpose of Report, Location, Method of Analysis, Findings, Recommendations, etc.). It gets me thinking and directs me so I don't accidentally miss a major feature. It also maintains a uniform look." I suspect that this professional also starts writing on day 1.

Benefits of Starting on Day 1

Imagine being involved, as a team leader or team member, in the parallel approach, that is, starting to write on day 1, and having the opportunity to contrast that experience with the sequential approach. You are likely to recognize and appreciate the following benefits of the parallel method:

- Fewer errors and omissions because it avoids the inevitable "last minute" rushed writing to meet a deadline. As suggested in Section 3.2.4, report deficiencies are likely to produce a "C" document that fails to describe fully and accurately what was "A" professional work. Delivering excellent work in a mediocre package means that everyone loses—those who did the work, their organization, and those who read the report.
- Effective communication within the project team. The evolving and frequently shared draft document keeps team members up to date and invites sharing of questions, suggestions, and concerns. This very specific intra-team communication is in addition to team meetings and various conversations.
- Effective communication with the project's primary beneficiary and stakeholders. Anyone outside of the project team who has an interest in the project receives a steady flow of information. They learn and can comment about assumptions made, information sources used, approaches taken, conclusions reached, alternatives considered, and likely recommendations—no one needs to be blindsided.

3.4.4 Structure Major Documents for Use by Varied Readers

Recall the first effective communication principle in Section 1.5, know your audience. Assuming you do, then structure documents so that various readers can find and read what they want and help you accomplish your purpose.

The 3-part structure shown in Figure 3.6 enables you to do that because it is reader-friendly. That structure is also writer-friendly because it provides you with an overall framework for your document-writing effort. Almost any document that we write—e.g., report, email or memorandum, professional article or paper, book—can be structured to include the three parts shown in Figure 3.6. So many of our writing efforts have those three parts, that is, an invitation-overview, substance-principal message, and support material.

Consider the audience for and content of each part of the structure for four types of documents—reports, emails or memoranda, professional articles or papers, and books.

Report

Part 1: Executive Summary In the engineering practice world, this one- or two-page part of a report meets the needs of busy executives such as corporate chief executive officers (CEOs), mayors, state and federal officials, and appointed

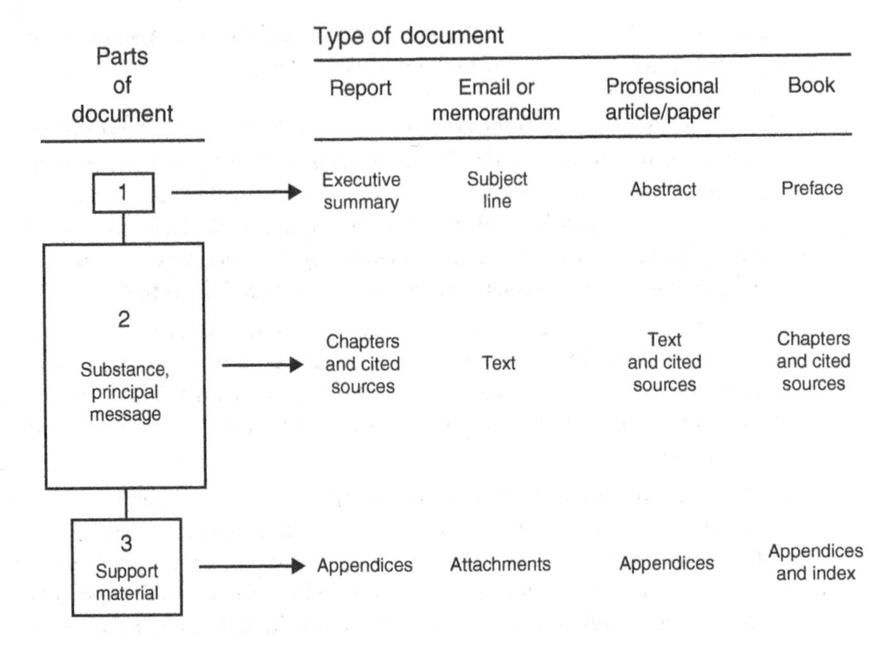

Figure 3.6 This 3-part document structure, which is applicable to a variety of document types, facilitates writer's writing and reader's reading.

board and elected council members. It concisely provides these decision makers with background, briefly describes methodology, summarizes conclusions, and focuses on recommendations with costs and a timetable. In academia, the document might be a final report about a capstone course project, where the executive summary is primarily for the top executives of the sponsoring organizations.

Part 2: Chapters and Cited Sources Directors of public works, chief engineers in industries, and other middle- and upper-level professionals and managers in the public and business sectors typically appreciate this much larger and very detailed part of a report. They tend to value a comprehensive description of a project from beginning to end, where "end" includes a list of cited sources. Part 2 is usually arranged in chapters, each of which has headings and subheadings, and concludes with that source list. For capstone courses, mainly faculty members involved in the course would read Part 2.

Part 3: Appendices Referenced from the text, each appendix is self-contained and includes detailed and voluminous material that appeals to a very small part of the total audience—a few scientific, technical, or other experts.

In the context of a capstone course, some appendices may be of interest to faculty and students, in various department across the campus, who heard about or assisted with the project. As suggested by Figure 3.6, the total size of the appendices, measured in words, is usually much smaller than the report's Part 2, but exceptions could occur.

Email or Memorandum

Part 1: Subject Line A short and well-written subject line will quickly grab the attention of potential readers and convey the essence of an email or memorandum so that most of them will read the document. Look ahead at Section 3.4.20 for ideas on how to write short statements that attract attention.

Part 2: Text Retain a reader's attention by describing the purpose of the email or memorandum in the first sentence or first short paragraph. What are you trying to accomplish? Then elaborate, keeping in mind that you do not have to say all you want to say in Part 2 because you have the option of using Part 3, that is, attachments. Present the content using headings and subheadings—they enable readers to skim your document, confirm their interest in it, and, at least initially, selectively delve into parts of it. Conclude the text with a gentle reminder of what you are seeking and say thank you or otherwise offer your appreciation.

Part 3: Attachments One or more attachments could add value to your email or memorandum, especially if you have a large and diverse audience. Because you know individuals, you will be able to place some complex, detailed, and specialized data and information in attachments, each intended for one or a few message recipients.

Professional Article or Paper

Part 1: Abstract Articles and papers differ mainly by the manner in which editors and others review them for possible publication. Articles typically appear in magazines, such as *PE-The Magazine for Professional Engineers,* published by the National Society of Professional Engineers (NSPE). A potential author unilaterally submits a draft article to the editor or is invited to do so by the editor. The editor and author make the publication decision. For a paper, an editor and the author's peers review the draft and make a publish-or-not-publish decision.

An abstract is sort of like an executive summary but much shorter. It summarizes the article's or paper's purpose, process, findings, and recommendations. In other words, what was the challenge, how did you and others approach it, what did you learn, and what do you recommend?

Part 2: Text and Cited Sources Readers will expect a detailed description of the aforementioned purpose, process, findings, and recommendations. Articles and papers usually use headings, and often subheadings, to organize and present content for the reader's benefit, similar to how a report or book uses chapters. Articles sometimes cite sources, while papers always do.

Part 3: Appendices Specialized data and information, likely to be of interest to a small fraction of readers, usually appear in appendices at the end of an article or paper.

Book

Part 1: Preface Nonfiction books, like those often written by engineers, usually have a preface, such as the book you are reading. The preface typically describes the book's purpose, intended audience, organization, and content; ends with acknowledgments; and uses headings for each. If I am attracted by the cover of a nonfiction book, say in a bookstore or library, I skim the table of contents and the preface; the two combine to become a mini-version of the book and help me decide if I want to read all or some of it.

Part 2: Chapters and Cited Sources Chapters traditionally give structure to Part 2, the book's principal content. Besides text, chapters often include supporting tables and figures, and each chapter ends with a cited source list. Sometimes a single and large cited source list appears at the end of all chapters or even later in Part 3.

Part 3: Appendices and Index As with reports, articles, and papers, a book's appendices include specialized data and information that are likely to be of interest to a small fraction of readers. Part 3 should also include an index, which enables readers to search for specific topics, and Part 3 may include abbreviations and glossary appendices. Clearly, appendices and an index help to make the book even more reader-friendly.

Let's use this book as an example of the relative size of the three parts. The preface and contents account for 3% of the book, the chapters and author bio account for 84%, and the appendices and index account for 13%.

***

I saw the need for the preceding 3-part document structure decades ago while working as a project manager at a consulting engineering firm. Because company management was not satisfied with report quality, they hired a retired high school English teacher to fix the situation. After reading or skimming a few of my documents, he said that I was writing reports for myself, not for readers. He explained his observation this way: I wrote chronological descriptions of what we did on a project. Readers had to wade through the entire document to find what they wanted. The 3-part structure is much more reader-friendly—it tells readers where to look for what they want.

The former teacher also helped us produce a company-style guide, like those discussed earlier in Section 3.4.1. Some initial skepticism changed to appreciation as the evolving guide proved to be valuable across the company.

Suggestion: Assume you are beginning a writing project; it may be a large document like a report or a small one, such as an email or memorandum. You profile your audience, articulate your purpose, and assemble some content. Then you apply the 3-part structure by beginning to think about where each item of content will ultimately go. You will have the framework for your document and be on the way to producing a reader-friendly result. Your readers won't have to wade through everything, like readers used to have to wade through my reports.

3.4.5 Create Bulleted and Ordered Lists

Assume you, a practicing engineer, draft a paragraph in which you list the three reasons you made a decision or indicate the five steps you used to analyze a failed process. Instead of the paragraph, consider using a bulleted list for the three reasons or a numbered list for the five steps. I offer this suggestion because lists highlight the reasons or steps and create some welcome variety and white space for the reader, as described further in Section 3.4.6.

For an example of a bulleted list, look back at Section 3.4.3. The use of neutral symbols—in this case, bullets—means that the three items are not necessarily sequential or prioritized. For consistency purposes, each bulleted item begins with a noun or nouns, that is, the name(s) of something. In bulleted lists, each entry could begin with a verb. The point is: Be consistent.

Refer back to Section 3.4.2 for a numbered list of seven items—they are numbered because the order is essential. In this case, each item begins with a verb. However, you could use nouns in a numbered list if the idea is to indicate decreasing priority or importance.

For both types of lists, use punctuation (e.g., period or comma) at the end of each item only when one or more of the listed items includes more than one sentence. For a no-punctuation situation, see the three-item bulleted list in Section 3.2.2.

3.4.6 Provide Reader-Friendly Features

Besides considering the 3-part structure for documents and creating bulleted and numbered lists, consider some more ways to make all of your documents reader-friendly, that is, attractive and easy to read. A mass of words presented in long paragraphs on multiple pages intimidates most potential readers. Avoid that mass mess by thoughtfully designing and constructing your document in the spirit of this observation by the Spanish writer Enrique Jardiel Poncela: "When something can be read without effort, great effort has gone into its writing."

Consider the following examples of reader-friendly ways to attract, engage, and retain your readers, largely regardless of the kind of document you are preparing:

- **Table of contents** that includes headings and subheadings, as is the case with this book.
- **Headings and subheadings,** as used throughout this book. Numbering all first-order, second-order, and higher-order headings enables referencing another section from any given section. Headings, not necessarily numbered, are applicable to other documents such as memoranda, letters, and emails.
- **Paragraphs,** as noted in the *Elements of Style*, hold text together. This results from us partitioning our subject into topics, "each of which should be dealt with in a paragraph." That book goes on to suggest, "As a rule, begin each paragraph either with a sentence that suggests the topic or with a sentence that helps the transition" [5].
- **Visuals,** such as graphs, photographs, line drawings, flow charts, and tables. This book contains 30 visuals. Chapter 5 explains why vision is our most powerful sense and describes how to prepare visuals that leverage that sense.
- **Word processing capabilities** such as different fonts, boldface, underlining, and color. This bulleted list uses boldface to identify user-friendly techniques. If you use color, anticipate how the colored items will appear when copied or printed in black and white. That is, would the intended emphasis be lost, or have you also used a different font or underlining?
- **Tabs,** attached to the right side of pages that begin a chapter, section, appendix, or other part of a document, can be useful in certain situations. The tabs might carry the name of a section or the capital letter indicating an appendix. Frequent users of documents, such as manuals that contain a large number of frequently accessed chapters, sections, or appendices, appreciate tabs.
- **List of abbreviations,** as illustrated by this book's alphabetized list in Appendix A. In any chapter, the first time a word that is abbreviated appears, spell it, and then immediately follow with the abbreviation in parentheses. From then on, use the abbreviation in that chapter.
- **Glossary,** that is, an alphabetized list of words that may not be widely known to the document's intended readers, along with their definitions. This mini-dictionary typically appears as an appendix.
- **Index,** which is essential for most nonfiction books and suggested for manuals and other large documents. I, as part of my initial evaluation of

a nonfiction book, skim the index, which is typically the last item in the book, because the presence of an index tells me that the author(s) had serious readers in mind.

- **White space,** which appears often in this book, because I proactively use many user-friendly features like the preceding. They automatically create white space around them, which readers tend to find attractive.

Reader-friendly documents, ranging in size from emails to books, have aesthetic appeal. Try this. Find a document prepared with some user-friendly features like those described above. Choose another document with little or no such features. Lay them side by side on your desk and then stand up and look at them, but don't read them. Turn some pages while forming an overall impression. Admittedly subjective, my guess is you will find the user-friendly document most appealing because of its visual variety, including restful white space, and its ease of use. Potential readers must notice and read your document for you to accomplish your objective. Reader-friendly features attract attention and stimulate interest.

3.4.7 Write in Mostly Active, Rather Than Passive, Voice

Features of Active and Passive Voice When Writing

"Active writing, or writing in the active voice, is direct, rigorous, definite, and shorter. Active voice typically uses a simple subject-verb-object structure. In contrast, passive writing, or writing in the passive voice, is indirect, tame, and sometimes indefinite" [2].

After reading those very different active and passive voice definitions, you may wonder why anyone would ever write in the passive voice, especially when considering reader preferences. Engineers and others write often in the passive voice for various reasons, including that they are simply not aware of the two voices. Therefore, they unknowingly produce mediocre results.

Consider this example of a passive voice sentence: "It was determined from laboratory analysis, that the manufactured chemical did not meet company standards." The example sentence begins with "it," which is a common indicator of passive writing. "It" has a "hidden antecedent" in that we do not know what preceding noun is referenced [14]. Skim some samples of your writing, and if you find many sentences starting with "it" and lacking obvious antecedents, you are using the passive voice.

We can rewrite the example passive voice sentence to create an active voice version as follows: "Laboratory analysis determined that the manufactured chemical did not meet state standards." Note the simple subject-verb-object structure. Writer and editor Patricia T. O'Conner explains the essence of the verb in the active voice this way: "An active verb makes somebody or something responsible for an action" [15]. In other words, the active voice indicates whom or what is doing whatever is being done.

Table 3.1 Readers benefit from the active voice because it produces stronger and shorter sentences.

Passive voice	Active voice
The reason she studied engineering was that she enjoyed creating things. (11)	She studied engineering because she enjoyed creating. (7)
Shortly after he offered his very negative comments about the design, he said he regretted his choice of words. (19)	He quickly apologized for his overly critical comments. (8)
There were a great number of soil boring holes all over the proposed construction site. (15)	Soil boring holes covered the proposed construction site. (8)
My first solo sailing will always be remembered by me. (10)	I will always remember my first solo sailing. (8)

Source: Adapted from Strunk and White [5].

The active voice version has 12 words, whereas the passive voice version has 15 words. Typically, the active voice requires fewer words. Therefore, if you strive to use the active voice, your readers will benefit from enhanced clarity and less text.

Consider the examples in Table 3.1, adapted from Strunk and White [5], which show how the active voice adds clarity and strength and reduces text. Parentheses indicate the number of words.

Note, in the right-hand column, how the active voice versions use the easily understood noun-verb-object structure and are significantly shorter. According to Strunk and White, ". . .when a sentence is made stronger, it usually becomes shorter. Thus brevity is a by-product of vigor."

Use Vigorous Verbs

Let's drill deeper into the active voice by considering the verbs it uses. Author and teacher Robert Greenman states, "Nothing animates and invigorates one's writing or spoken expression more than well-chosen verbs, the heartbeat of every sentence." He goes on to say that nothing can be, do, or happen without a verb [16].

Getting more specific, Greenman says the most active verbs tend to include those that are also nouns or derived from nouns. For example, **engineer** is a noun when used like this: John F. Stevens, an **engineer**, invigorated the failing Panama Canal project by viewing it as a railroad project.

We can use **engineer** to derive a vivid verb and apply it like this: Stevens, drawing on his railroad experience, **engineered** a solution to the failing Panama Canal project [11]. In the last sentence, notice how the verb **engineered** is more vivid and related to engineering than some weak verbs, such as found, implemented, or provided.

Consider an example using the noun **license** as the basis for a verb. States **license** individuals as professional engineers (PEs) after they earn an accredited engineering degree, satisfy experience requirements, and pass the fundamentals of engineering (FE) and principles and practice of engineering (PE) examinations.

You too, when writing or speaking, can use nouns, some drawn from your specialty, either as is or to derive a verb, to animate your written or spoken message. Your audiences may connect with science-technical-engineering nouns such as crash, creation, design, failure, diagnosis, innovation, model, and simulation. Any of them could be the basis for an active verb.

Situations Where the Passive Voice is More Appropriate

Experience reveals special situations where the passive voice accurately and more empathetically describes an action or event and, therefore, is preferable. For example, recall the Section 1.1.2 description of two Boeing 737 MAX 8 crashes. This passive voice sentence begins the discussion of the tragedy: "In October 2018 and March 2019, 346 people were killed when two Boeing 737 MAX 8 aircraft crashed in a similar manner, nose down at over 500 miles per hour." This sentence is accurate.

In contrast, an active voice version, would read "In October 2018 and March 2019, two Boeing 737 MAX 8 aircraft crashed in a similar manner, nose down at over 500 miles per hour, killing 346 people." This active voice sentence is also accurate. However, it puts too much emphasis on the aircraft and gives too little attention to deceased individuals and their survivors. Therefore, although I often write in the active voice, I used the passive voice in this special case. The passive version seems more appropriate because it mentions fatalities first, which focuses the reader on lost lives and the impact on survivors. Be alert for possible exceptions to favoring the active voice [14].

You may also decide to occasionally use the passive voice to add variety to your text. Reading your text out loud, as suggested later in Section 3.4.22, will help you detect the need for variety. Incidentally, when used in the spelling and grammar review mode, some word processors nicely flag passive voice sentences for you.

3.4.8 Pick Your Person

When we write or speak, the use of first, second, and third person indicates the point of view, that is, respectively, who is narrating the story, who is receiving the message, or what happened regardless of who was part of the story [17, 18]. Consider the following examples of the three persons, the three points of view:

- **First person:** "We wrote the software that enables the 3D printer to produce the prosthetic device." The storyteller is part of the story. "We" and "I" are common first-person pronouns.

- **Second person:** "You wrote the software that enables the 3D printer to produce the prosthetic device." The reader of the story is part of the story. "You" is a common indicator of the second-person point of view.
- **Third person:** "They wrote the software that enables the 3D printer to produce the prosthetic device." Some unknown others are the players in the story. "They," "she," and "he" are common third-person pronouns.

Having introduced the three points of view, we naturally ask what is their value. For starters, consider a writing experience that illustrates the effective use of first and second person. I prepared a proposal for an engineering textbook and sent it to a potential publisher—the proposal package included an example draft chapter written in third person. A publisher representative responded favorably, with one exception. He suggested that I write the book in mostly first and second voices because that would enable me to speak informally and conversationally with students. I did as suggested, the textbook was published [19], and I subsequently wrote other books, including this one, using, mostly first and second person.

Use of second person can be effective when writing "how to" documents like manuals and guides. By frequently writing "you," you speak directly to the reader.

Third-person writing is often required in journals that publish peer-reviewed papers. The idea is to encourage writer and reader objectivity by encouraging the writer to present the facts and not share his or her views.

Sometimes writers use an abrupt shift from one point of view to another as a means of seizing reader's attention. For example, you, as a PE might be the principal author of a report describing the cause of a problem and ending with your firm's recommendations for solving it. You write everything prior to the recommendations in third person, and then, at the start of the recommendations, you switch suddenly to first person by writing "We recommend. . ." and listing the recommendations. By changing to first person, you assure reader attention and remind readers that the recommendations reflect your objective engineering analysis plus your and your team's hard-earned experience and judgment.

3.4.9 Write Critical or Easy Parts First

When I am writing a book, I draft the preface first, as I did for this book, because it compels me to articulate the book's purpose, identify the intended audience(s), describe the content, and outline how the book is organized. I even begin drafting the acknowledgments section of the preface because, if I have decided to write a book, the advice or work products of one or a few individuals influenced me—they should be acknowledged.

A glossary, which would eventually appear as an appendix at the end of a report, would be an excellent first-written product when a team is working in a new area of technology. Drafting, reviewing, discussing, and refining the critical glossary will align team members based on their understanding of some new terminology and get everyone on the same page.

You may, as the designated writer of all or portions of a document, be most familiar with or enthusiastic about certain aspects of the topic. Maybe it's computer modeling of systems, data collection and analysis, robotics, or 3D printing. Then go after the low-hanging fruit first; write about your favored feature so that you get some words on paper and begin to beat down writer's block. We do not have to write about topics in the same order that they occurred or will appear in the final document.

3.4.10 Use Metaphors and Similes

The dictionary [20] defines these two powerful figures of speech as follows:

- **Metaphor:** "a figure of speech in which a word or phrase literally denoting one kind of object or idea is used in place of another to suggest a likeness or analogy between them (as in *drowning in money*)."
- **Simile:** "a figure of speech comparing two unlike things that is often introduced by *like* or *as* (as in *cheeks like roses*)."

We see the communication power of metaphors and similes in that both help us understand something that may be new or unfamiliar to us by comparing or identifying it with something that is familiar to us. The metaphor is the most extreme or daring of the two figures of speech in that it "consists of images connected to something they literally cannot be" [21].

Examples of Metaphors

- Jordan earned a mountain of praise for his efforts to make the project profitable.
- The essay Jerrie submitted in the competition nailed the first prize for her.
- An engineering license propels you to a successful and satisfying career.
- As you read any text, I suspect you will now be more likely to flag metaphors.
- Sales are frosting on the marketing cake.

Notice that in these examples, no one actually earned a mountain, nailed, propelled, flagged, or frosted anything. However, those eye-catching features strengthen the text and remind us that we don't always have to be strictly literal.

Examples of Similes

- Jordan earned what seemed like a mountain of praise for his efforts to make the project profitable.

- Earning an engineering baccalaureate degree and not taking the FE examination is like finally buying your dream car and not taking it for a drive.
- Editing for brevity is like pruning the text.

Note how similes use words such as "like" and "as" to explicitly connect two different entities.

3.4.11 Retain Readers with Transitions

As you draft text, every now and then, intentionally encourage readers to stay with you by using inviting transitions between some sentences and from one paragraph to the next. Strive to write text that, like a magnet, draws readers in and then compels them to stay. Consider three types of "magnetic" transitions.

The For Example Transition

Notice how the following text makes a statement using a three-legged stool metaphor. Then a "for example" transition appears to encourage the reader to read on within the paragraph.

A successful consulting engineering firm is a sturdy stool supported by leading, managing, and producing legs. While the firm might survive temporarily balancing on two legs, long-term success requires all three legs. **For example,** a leaderless firm might do well for several years—live off its managing and producing legs—but eventually topple because it lacks the ability to see and act on changes in client needs.

Most readers welcome examples because they aid understanding. Therefore, occasionally use example transitions to meet that reader's need and retain them so they benefit from your message.

The Question Transition

The following paragraph begins with simple statements and then asks a question. It seems strange for the writer to ask the reader a question that the reader cannot answer for the writer. That is the purpose of the question transition, asking a question is a rhetorical technique used to retain attention. Having assured your attention, the writer makes a key point: Big relationships begin with small acts.

Some individuals regularly fail to keep small promises. For example, they promise to send you an article but don't. Some might label breaking promises about "little" things harmless oversight. **If part of a pattern, are small broken promises harmless?** Some people may tolerate an engineer with a reputation for breaking small promises because, while annoying, the consequences for all of them seem small. In contrast, accumulated failures to keep small promises may prevent development of trusted relationships with other co-workers, clients, customers, elected officials, regulators, and others on major issues. Those individuals may conclude that, if you cannot be trusted to keep small promises then the same may apply, with serious consequences, for large promises.

When writing, be alert to the possibility of asking and then answering a rhetorical question as a means of maintaining the reader's attention.

The Single-Word Transition

The common words "and," "because," "but," "however," "if," and "whether" can stimulate your reader to keep reading. Consider an example using "however."

Engineering builds on a scientific and technical foundation. Given the importance of that foundation, the communication deficiencies of a highly intelligent and scientifically and technically competent engineer would seem to be irrelevant. **However,** "the most exciting vision, the most elegant solution, or the most creative design are all for naught unless they are effectively communicated to others" [2].

Note how the text draws a tentative conclusion and then, using "however," contradicts it and retains a reader's attention.

3.4.12 Punctuate Purposefully

Merriam-Webster defines punctuation as "the act or practice of inserting standardized marks or signs in written matter to clarify the meaning and separate structural units." An alphabetized list of example punctuation marks taken from a larger list in that dictionary follows: apostrophe, braces, brackets, comma, dash, ellipsis, exclamation point, hyphen, parentheses, period, question mark, quotation marks, and semicolon [22].

Shifting from the definition of punctuation to its purpose, my favorite description of the latter is that by author Lynne Truss, "Punctuation herds words together, keeps others apart" [23]. Her book, *Eats Shoots & Leaves*, uses a humorous approach to providing practical punctuation advice. Humorous punctuation may sound like an oxymoron; however, Truss's book is both helpful and humorous. For example, for those of us who struggle with apostrophes, she writes: "For every apostrophe omitted from an it's, there is an extra one put into its. The number of apostrophes in circulation remains constant."

Continuing with punctuation's purpose, consider this statement: "Construction is about to stop because of the snow and if the rebar load doesn't arrive soon it will have to be temporarily returned to the supplier." Will the snow or rebar be returned to the supplier? We can clarify the text—an example of a run-on sentence—by breaking it into two sentences and adding a comma like this: "Construction is about to stop because of the snow. If the rebar load doesn't arrive soon, it will have to be temporarily returned to the supplier" [24].

Run-on sentences, like the following example, lack clarity, "Judy knows how to perform calculations she never seems to make errors." Punctuation can come to the rescue, with the simple insertion of a semicolon as follows, "Judy knows how to perform calculations; she never seems to make errors." Another approach would be to insert the word "and" instead of the semicolon.

Recall the Section 2.4.3 discussion of body language. Then consider journalist Russell Baker's thoughts about the function of punctuation in effective writing. "In writing, punctuation plays the role of body language. It helps readers hear the way you want to be heard."

To assist you in providing proper punctuation for your readers, I prepared Appendix E. It addresses this commonly used subset of punctuation marks: apostrophe, comma, semicolon, hyphen, dash, italics, and ellipsis.

3.4.13 Tell True Personal Stories

Because true personal stories combine facts and feelings—speak to head and heart—they tend to attract curious readers who then typically understand and often remember the story's messages. Notice that we are discussing true, not hypothetical, stories. While a writer can use both types, true personal stories have more credibility—favor them.

Individual stories don't prove anything because they usually describe one isolated event or incident. The principal value of a true personal story is that you can use it to illustrate, in a cognitive and emotional manner, some idea or principle of potential value.

To illustrate how true stories attract readers and illustrate an idea or principle, let's briefly review the seven true personal stories I already told in this chapter:

- **Section 3.2.3:** Writing to the president of my company to present a new idea when, in retrospect, I should have first spoken to him
- **Section 3.3:** Misreading the title of a book, which led me to stress the idea of writing to learn
- **Section 3.4.1:** Seeing inconsistency in draft texts received from project team members and, for the sake of report consistency, leading preparation of a project-specific style guide
- **Section 3.4.2:** Being challenged with the need to prepare a cold call presentation and discovering that mind mapping generates rich content
- **Section 3.4.4:** Learning that I was writing reports for myself and that what I now call the 3-part document structure would help me focus on the audience
- **Section 3.4.8:** Deciding to use major first and second person in writing a book because of the thoughtful suggestion of a book publisher representative
- **Section 3.4.9:** Writing the critical preface of this book first because it gave me the opportunity to get started—to think and write about the book's purpose, audience, content, and organization

Even if you are a young college student, you have stories to tell because of your personal successes and failures. Imagine how your story collection will grow in quantity and quality as you move through your career. It did for me, and I find them useful for teaching and learning. Look for writing and other opportunities to tell true personal stories.

Engineer Richard G. Weingardt, PE, said, "It's only a mistake if you don't learn from it." Writing a story about one of our mistakes will help you, me, and others learn from it and help advance whatever mission or cause we are writing about.

Let's take a short time out to reflect. In this and the previous three sections, we discussed metaphors and similes, transitions, punctuation, and stories. Some of these and similar writing techniques described in this chapter may seem trivial, ornamental, or worthless. According to two experienced authors [25]: "Facts are not boring, but you [or I] might be." I offer this chapter's writing advice so that your writing, especially when communicating facts, is not boring—instead, it is likely to attract, engage, retain, inform, and convince readers.

3.4.14 Craft Informative Titles for Figures and Tables

Go back to Figure 3.4 in Section 3.4.2, which is titled with this declarative sentence: "This mind map generated content ideas for a cold calls presentation." I could have titled it "Example of a mind map," which would have been accurate but not as informative as the actual title.

Consider Table 3.1 in Section 3.4.7. I could have accurately titled it "Active and passive voices" but, instead, used "Readers benefit from the active voice because it produces stronger and shorter sentences." Again, more clarity for the reader.

When writing, we provide figures and tables to illustrate or further explain our text. Get maximum value out of any given figure or table by using a title that states, in a declarative sentence, what it illustrates or explains, as shown by the preceding two examples. This is one of the few places in this book where I am advocating using more words than necessary because, when titling figures and tables, the additional words needed to form declarative sentences are more likely to benefit readers. I practice what I preach throughout this book.

While we are discussing titling figures and tables, consider a thought about numbering them in documents with two or more chapters or sections. I suggest restarting the numbering in each chapter or section and including the chapter or section number. This approach, which is illustrated in the just-discussed Figure 3.4 and Table 3.1, and used throughout book, including appendices, simplifies the almost inevitable addition and removal of figures and tables during the drafting and revising process. Adding or removing a figure or table affects the numbering only in its chapter or section, not all subsequent chapters or sections.

3.4.15 Avoid Liability

As you move ahead in engineering practice with a consulting engineering firm, someone is likely to ask you to draft portions of a proposal. The document will describe how your firm proposes to provide services to an existing or potential new client. Alternatively, you may be on the engineering staff at a large manufacturing company and tasked to write part of an agreement that describes how your company will design and manufacture a product for a customer.

Why We Need to Watch Our Language

You and others must carefully craft these frequently written documents so they clearly and unambiguously describe what you and others will do and not do. That desired clarity depends on choosing appropriate words. Ambiguity in your firm's proposals, agreements, and similar documents could result in your company being liable if a conflict occurs. Liability means that your firm must pay or otherwise compensate a client, customer, or other person or entity.

Consider a hypothetical example. The aforementioned proposal, finally written by you and other members of your consulting firm, included this statement: "Gather and review all existing information." You wrote that sentence to convey your firm's thoroughness. The client accepted the proposal and then executives of your firm and the client signed an agreement. The quoted sentence appears in the agreement.

As the project nears completion, a member of the client's team notes that your firm did not discover and use an important report, which may have bearing on the project's findings and recommendations. This results in some conflict during which the client's representative reminds you and your colleagues about the "gather and review all existing information" promise. Your firm is liable, not monetarily, but in the sense that you are obligated to review the now-discovered report, review it, and determine if it affects your findings and recommendations. That obligatory additional effort could be costly and reduce the project's profit.

The preceding problem would have been less severe if, instead of the "all" statement, the original proposal and then the agreement had used this statement: "Collect and review as needed readily available information."

Words matter in the liability world, even a little three-letter word like "all." If you are drafting text for a proposal, agreement, or similar document do a word search on all. "I'm not suggesting watching our language in contracts and agreements as a way to 'put one over' on a client, mislead an owner, or 'slip one by' a customer. I am suggesting watching our language in order to communicate effectively" [26].

Table 3.2 The left column shows typical problematic statements, and the right column offers suggested replacements.

Problematic statement	Suggested replacement
Gather and review all existing information	Collect and review, as needed, readily available information
Prepare summaries of all project meetings	Prepare summaries of monthly client and consultant meetings
Visit the project site periodically	Visit the project site at least once each calendar week
Assure that the public, via three public meetings, will accept the study's recommendations	Assist the client, via three public meetings, in urging the public to accept the study's recommendations
Inspect and supervise work at the project site	Observe and report on work at the project site
Ensure that the contractors perform as per the approved plans and specifications	Observe the contractor's work relative to the plans and specifications and report discrepancies to the owner
Obtain permits to construct the water treatment plant	Assist with obtaining permits to construct the water treatment plant

Source: Adapted from Refs. [27, 28].

How to Watch Our Language

Having described why we need to watch our language, look to Table 3.2 for ways to do so. Well-intentioned individuals could have written entries in the left column, but they are problematic. The right column offers suggested replacements that convey a similar intent but add clarity and reduce the likelihood of causing conflict or incurring liability.

3.4.16 Give Credit

Years ago, while reading an article, I encountered a familiar paragraph—familiar because it was taken word-for-word from an article I had written. Because of failure to acknowledge my efforts, I contacted the author to explain my concern. He blamed his assistant. Frankly, I felt violated. Using my intellectual and creative products without credit is like breaking into my home and stealing my flat-screen television. Theft is theft.

Plagiarism, that is, using the work products of others without crediting them, has always been tempting to engineers and practiced by a few at risk to their reputations. The temptation and negative consequences are greater today because we operate in a digital world. The internet enables potential plagiarists to find, more easily, content to steal. Because we live in an increasingly digital world, plagiarists' actions will be out there in cyberspace forever and could come back to haunt them.

When our writing efforts use text, visuals, ideas, and information created by others, let's give them credit and expect others to do the same. An anonymous person, apparently a book author, who reviewed the draft of this book said,

"The references are superb! Maybe even too many!" I value those comments because they recognize my effort to give credit where credit is due.

We have many options for giving credit. For example, as illustrated in this book, I use bracketed numbers in the text to indicate that I have drawn on the work of one or more other writers or speakers. Those sources are keyed to a numbered list in the References section at the chapter's end. This is the Vancouver referencing system.

It contrasts with the Harvard system, in which authors' names and years of publication appear within parentheses in the text (e.g., Smith 2018). They link to an alphabetical list of authors in the chapter's References section. There are other options. Use what pleases you unless required, for example, a journal or magazine, which is common, to apply a particular referencing system.

If you write, or help to write, a major document like a project report, a manual, or a book, consider including acknowledgments. For an example, see the acknowledgments section in the Preface of this book. This kind of section offers a gracious and explicit way to thank individuals and organizations who helped you and/or your team in ways that go beyond formally citing sources. You may want to add a caveat that says, while you appreciate the help of others, you and/or your team are responsible for your document's findings and recommendations. Refer again to this book's Acknowledgments section for an example caveat.

3.4.17 Minimize Euphemisms

The Merriam-Webster Dictionary defines a euphemism as "substitution of an agreeable or inoffensive expression for one that may offend or suggest something unpleasant" [29]. We engineers don't want to offend our readers but we also want, in the interest of being helpful to those we serve, to be truthful and specific. Accordingly, we will occasionally find ourselves in a quandary and have to call on our best judgment. In the spirit of understanding euphemisms, consider the hypothetical and somewhat humorous examples in Table 3.3.

Table 3.3 Euphemisms in the left column become reality in the right column.

What someone said	What they probably meant
Conducted thorough research	Googled for ten minutes
For your information	We don't want to deal with this; you figure it out
Program	Any task that can't be completed in one day
Reliable source	The Uber driver who drove me to your office
Give us the benefit of your thinking	We will probably ignore what you say unless you agree with us
Your request is in process	It's hopelessly wrapped up in red tape
My advice is based on a careful analysis	I made some quick calculations on the back of a restaurant placemat

Source: Adapted from Loeffelbein [30].

Table 3.4 When first mentioning an object in written text, be specific.

Not specific	Specific
Large clarifier	60-foot diameter, 15-foot deep clarifier
Structural deterioration	Failed beam
Powerful engine	3.5 liter 375 hp engine
Costly wind tunnel	$1.8 million wind tunnel
Fast passenger train	190 mph passenger train

Source: Adapted from Berthouex [32].

3.4.18 Strive for Specificity

The dictionary defines specificity as "the quality or condition of being specific" [31]. We engineers often write about products, structures, facilities, systems, processes, or parts of them. We often provide drawings, diagrams, photographs, and other visuals to provide specifics. However, we can also provide the reader with some specificity in the text, especially the first time we mention an object.

As suggested in Table 3.4, avoid a general description when first writing about something—use specificity to engage and inform the reader. For example, when first mentioned, the large clarifier is described as the "60-foot diameter, 15-foot deep clarifier." After that, writing just "clarifier" or "large clarifier" would be adequate, and introducing a visual would be appropriate.

3.4.19 Use History to Support Your Purpose

According to Harry S. Truman, America's 33rd President, "There is nothing new in the world except the history you do not know." His observation suggests that history tends to repeat itself—the good and the bad. When moving forward, you can strengthen your writing and speaking by using history for insights into what to do and what not to do.

Consider three hypothetical examples that draw on historic events:

- **Contemplating an unorthodox solution**: You are suggesting that your team take an unorthodox approach to solving a challenging engineering problem—and you are getting pushback. Reinforce your argument by describing how the struggling Panama Canal project was rejuvenated and brought to completion when John F. Stevens, the new engineer in charge, determined that the principal obstacle was not excavating but removing excavated material. The solution: View and complete the project by managing it as primarily a railroad project, not an excavation project (Section 2.3.5). You tell your team, via writing and speaking, that take-a-new-approach story.
- **Creating a unique project report**: Your senior project team is determined to produce the most thorough and engaging final product, especially

the oral presentation. How could your team produce a stellar end-of-the-project presentation to students, faculty, and sponsors? To stimulate creative thinking about ways to set your final presentation apart from the others, you share with team members the Taco Bell rebuild-the-restaurant-in-48-hours story (Section 2.3.5). It generated large and long-lasting attention for the company.

- **Urging a team to persist**: Your student or practitioner team took on a major challenge, which has proven to be even more demanding than expected, and you continue to encounter setbacks. Using memoranda and thoughts offered at team meetings, you urge persistence. You tell stories about how history rewards tenacity, including some of the ten examples from within and outside of engineering listed in Section 1.3.4.

Another thought about history offered by British writer Thomas Carlyle is, "The true past departs not; no truth or goodness realized by man ever dies, or can die; but all is still here, and, recognized or not, lives through endless change." History can be one of your greatest writing and speaking resources.

3.4.20 Title to Attract Attention

I placed this titling tip near the end of the writing advice section because it is one of the last things to complete prior to editing. However, think about the title occasionally as you draft an email, report, or other document so that you gradually create a title that attracts attention and suggests your purpose.

If you use many of the writing tips offered in this chapter, I am confident that your writing will gradually improve. You will increasingly produce more effective documents, ranging from short emails to major reports. Of course, each well-written document must be read in order for you to accomplish your writing goal—your purpose.

Anticipate Competition

Regardless of how you send your document, upon arrival, it will compete with many other documents. The email you are about to transmit to a key decision-maker or source of critical information will compete with many other emails received today by that person. Similarly, the major report you mailed yesterday, via the U.S. postal service, will compete with other envelopes and packages, most of which will be quickly opened, prioritized, or discarded.

What's my point? Essentially, everything we write vies for the attention of our intended recipient(s). Fortunately, we can give every document we write a title. An email's title appears on the subject line. A report's cover displays its title, which may also appear in a transmittal letter or memorandum. Therefore, we should craft document titles that serve as magnets in that they attract the intended reader(s)—and suggest your purpose.

You may argue that your writing should be valued on its merits, not on a "catchy" title. You are correct. However, to be judged, it must be noticed and read.

Examples of Effective Titles
Let's use emails as examples of titles intended to attract readers. These subject line entries appeared on emails I sent or received.

- Books and books' people
- Frustration
- Let's do it!
- Problem solvers and beyond
- Thanks for the opportunity

Alliteration, that is, repetition of a sound in a title, will attract some readers to an article, report, book, or other document. Some examples:

- Force or Finesse?—Title of journal article [33].
- Preparing, Presenting, and Publishing Professional Papers—Lesson in a book of lessons [34].
- *Engineering's Public-Protection Predicament: Reform Education and Licensure for a Safer Society*—Title of a book [35].

Consider using a two-part title for reports, books, and other larger documents. The first part introduces the subject, and the typically longer second part provides a little more detail. Separate the two parts with a colon. The two-part title method gives you, the author, two opportunities to engage a potential reader. The bulleted book item immediately preceding this paragraph illustrates this two-part title approach. Other examples follow:

- *Visual Thinking: The Hidden Gifts of People Who Think in Pictures, Patterns, and Abstractions*—a book [36].
- *Engineering Your Future: The Professional Practice of Engineering*—a book [19].

To reiterate, titles like the preceding examples hopefully attract readers but don't necessarily inform—that's the document's purpose. Clearly, your, my, or others' reaction to any title will be subjective. In spite of our good intentions, a title we create may offend, disturb, or repel some individuals. However, like hockey great Wayne Gretsky said, "You miss all the shots you don't take."

3.4.21 Prune the Text

In the plant world, pruning means removing superfluous matter to encourage fruitful growth. The same principle applies in effective writing. As you work through many draft and revision cycles, prune superfluous words and make other changes so that your written document yields fruitful results, that is, achieves your purpose. More bluntly, don't burden the readers with excess growth.

Another way of viewing pruning is to recognize that engineers, and those they serve, strive to design and manufacture or construct efficient products, structures, facilities, systems, and processes. Listen to Strunk and White: "A sentence should contain no unnecessary words, a paragraph no unnecessary sentences, for the same reason that a drawing should have no unnecessary lines and a machine no unnecessary parts" [5]. Appreciation of efficiency suggest that engineers, beginning as students, would value efficient writing, consisting mostly of essentials, and use pruning as one way to achieve it.

Stephen King, a successful author of over 50 books, uses a pruning formula that he learned in high school: Second draft equals first draft minus ten percent [37]. Informed by experience, most of us can readily reduce the word count by ten percent during our first revision. Consider some specific ways to prune your writing.

Delete Useless Words

Your or my first draft of a paragraph for use in a short email or a long report will, in spite of our good intentions, include useless words. Look for and eliminate them inspired by the following examples, where bold italics identify the unnecessary words each of which can be simply dropped [14, 38].

- We recommend using diesel for fuel ***purposes***
- The red-***colored*** comments were added by Mary
- Confirm that the tank is ***completely*** full before opening the valve
- Enter your PIN ***number*** to gain access to the files
- Please review the project's ***past*** history to determine the cause of the failure

By sharing the preceding examples of useless words, I hope that you will be even more inclined to search for and delete them from your drafts.

Consider some thoughts on the effective use of the word currently. Engineers and others often use this word in situations where it adds no value. For example, if someone asked you what year you are in at the university, why would you say I am currently a sophomore? Instead, simply say, I am a sophomore. "Currently" adds nothing in this situation—it's offering an unneeded qualification. When you are about to say or write "currently," ask yourself if it is needed.

We encounter circumstances, typically in rapidly changing or fluctuating situations, where the word "currently" aids accurate communication. Assume you are a member of a project team, have several assigned tasks, and move from task to task in response to team member needs. If the project manager asks you what you are working on, you might helpfully say that I am currently working on the data analysis task. Given the project manager's coordination responsibility, knowing what you are working on right now is likely to be helpful.

I enjoy writing to the point of finding some word combinations humorous. Consider these examples of word pairs in which the useless word might prompt a smile [14]: advance planning, totally destroyed, revert back, temporarily suspended, and illegal crime.

Replace Two or More Words with One Word

In addition to deleting useless words, we can also reduce text length and enhance clarity by replacing two or more words with one word or fewer words. Consider some examples where one carefully selected word replaces three to five words, with the replaced words shown in bold italics [32, 38].

- This treatment process ***exhibits the ability to*** meet state and federal standards. Replace with "can."
- ***Owing to the fact that*** pier scour occurred, we recommend reinforcing the foundation. Replace with "Because."
- The project team ***held a meeting*** to evaluate optional solutions. Replace with "met."
- Jack ***pointed to the fact*** that we need a decision by Monday. Replace with "noted."
- The firm ***lacked the ability to*** provide the requested services. Replace with "couldn't."

We conclude this discussion of replacing two or more words with one word or fewer words by illustrating how you might edit a paragraph of draft text to reduce the number of words and clarify the message [39]. Imagine that you just drafted a report that included the following 41-word sentence, and you are taking a second look at it:

"During the course of the study, we were able to determine that the chemical processing plant exhibits the ability to accommodate only up to one million gallons per day of flow owing to the fact that the pumps have that capacity."

While the message is clear—pumps limit plant capacity—the sentence could be much shorter for reader's convenience and for clarity. Here is the same 41-word sentence with certain words that caught your attention, highlighted in bold italics:

"During ***the course of*** the study, we ***were able to determine*** that the chemical processing plant ***exhibits the ability to*** accommodate only up to one million gallons per day ***of flow owing to the fact that*** the pumps have that capacity."

On thinking about your potential readers, you decide to:

- Delete ***the course of***
- Replace we ***were able to determine*** with "determined"

- Replace ***exhibits the ability to*** with "can"
- Replace ***of flow owing to the fact*** *that* with "because

Now you, and probably your readers, will be pleased with this resulting 26-word edited sentence: "During the study, we determined that the chemical processing plant can accommodate only up to one million gallons per day because the pumps have that capacity."

You reduced the length by 37% and greatly improved clarity. Amazing! Assuming you similarly edit your entire first draft, you will treat your readers to a great reading experience. The communicative engineer is an aggressive pruner and, as a result, harvests great results.

Getting rid of useless words, reducing use of "currently," or replacing two or more words with one word may seem trivial, a waste of time. However, when we read well-written documents and ponder the drafting and editing effort, the conclusion is likely to be that the writer consciously or subconsciously attended to a myriad of details, including pruning, and we enjoy the cumulative effect. As noted at the beginning of this chapter, "hard writing makes for easier reading."

3.4.22 Read Your Writing Out Loud

To introduce this topic, please read the following paragraph to yourself—not out loud:

The objective of the study is to identify all trends in the market for the company. The study will identify products that offer added opportunities to the company. In addition, the study will identify the requirements that must be met by the company [40].

The text makes sense in that it describes how the study will provide three benefits to the company. However, now read the paragraph out loud—hear the words in addition to thinking about them. You are likely to hear, if you did not already see, that the words "study," "identify," and "company" each appear three times. These repetitions within three sentences, even if not heard, will strike many of us as being objectionable or even unaesthetic.

Therefore, we could make the text more pleasing by revising it as follows:

The objective of the study is to identify trends in the company's market, products that it might add to its line, and the requirements imposed on the company by these products.

When read out loud, this version sounds more pleasing, largely because "study" and "identify" each appear once and "company" twice, once as a possessive and once as a noun. Furthermore, the revision reduces the text from 43 to 31 words without affecting the meaning. This text reduction provides another illustration of pruning.

Out loud reading of the draft text of your email or a report chapter provides results that are more aesthetic and reduce the amount of text. Other potential benefits include discovery of and an opportunity to rectify [2, 41]:

- Excessively long paragraphs or sentences
- Too many sentences of the same length
- Missing words
- Same word repeated soon
- Inadequate or excessive punctuation
- Misspelled words
- Excessive passive voice
- Ineffective person
- Inappropriate tone
- Confusing syntax
- Undefined terms
- Lack of specificity
- Poor transitions from topic to topic

Frankly, you or I may at times feel foolish reading our writing out loud. Small price to pay for a resulting large benefit to our readers. After writing, reading, re-writing, and re-reading all parts of this book, I read all of it out loud prior to submitting the manuscript to the publisher for copy editing. That effort helped me make many little improvements, with the cumulative effect of producing an even better book for you.

3.4.23 Arrange for Editing

Introduction to Editing

For use in this book, think of editing as a collaborative process that begins with a writer preparing a draft document. It could be any one of the many different types of documents mentioned in Section 3.1. The editor reads the draft and suggests corrections, condensations, and additions in areas such as spelling, grammar, punctuation, sentence structure, tone (e.g., formal or conversational), overall organization, headings and subheadings, figures, tables, content, relevance, and scientific and technical accuracy.

The writer responds by accepting some suggestions in whole or in part and rejecting others. Editing succeeds—produces a better result—because two or more different and engaged minds are more effective than one [42]. As observed by author Luc Sante, "Sometimes a glaring error that you motored blithely past a dozen times will become apparent only on the 13th read. Perhaps, rather than read your draft 13 times, you might want to hand off your draft to others" [43].

The 13 may be an exaggeration, but the underlying advice is correct. The surest way to produce writing errors is to be your own editor.

Some of us, maybe including you, want to go beyond just finding errors. We also want to find the most meaningful words, as nicely stated by fiction writer and humorist Mark Twain: "The difference between the right word and the almost right word is the difference between lightning and the lightning bug." When you begin to receive diminishing returns from your self-editing but want a better product, hand it off to an editor for polishing, including finding the right words.

Caution: Before handing your draft to someone for editing, always apply the editor built into your word processor—such as Word's spelling and grammar checker. Doing this spares your editor from having to make changes that you could easily have made and, therefore, enables your editor to focus on higher-level content and writing improvements. Note that Word's checker checks more than just spelling and grammar. For example, it flags passive voice, which, as suggested in Section 3.4.7, we should usually replace with active voice because it produces shorter and clearer text.

If you want to be even more helpful before transferring your draft to an editor, consider some of the following proofing ideas [2]:

- Examine headings and subheadings for consistency of style and fonts and for conformance with the table of contents
- Read the text out loud one more time
- Lay the text on your desk, stand up, and look at your product for its overall effect—might more subheadings, paragraphs, or white space improve the aesthetics?
- Take the draft to a new setting and read it there
- Stay consciously away from your draft text for an hour or a day, let your subconscious mind work on it, and then read one more time to learn what your subconscious suggests

Two Categories of Editors

Most engineers will eventually work with, or at least learn about, two types of editors—formal and informal. Formal editors carry some kind of editor title, such as chief editor, managing editor, or copy editor. They may work for a newspaper, magazine, journal, book publisher, or your employer. Some form or aspect of editing is their principal function [41]. Because of my writing of articles, peer-reviewed papers, reports, and books, I have worked with many formal editors. I say unequivocally that everyone added value to what I wrote.

The informal editor, in contrast to the formal editor, has no official title. This person may be your spouse or significant other, a friend or colleague, a professor, your boss or supervisor, someone you just met and shared an interest with, or anyone who you think would help you by reviewing a draft. My wife helps me by reading almost every document I write. She read this entire book once, and parts of it more than once. She comments most about spelling, grammar, punctuation, and tone and always adds value.

As a student or young practitioner committed to improving your writing knowledge, skills, and attitudes (KSA), informal editors will do most of your editing. With experience, friends and others will increasingly ask you to be an informal editor, and eventually, you are likely to work with formal editors.

Avoid Being Adversely Affected by Criticism

If you are overly sensitive—thin-skinned—about having your work criticized, get over it or your final writing will suffer. Editors may sense your sensitivity and not give you their best work. Ask for strong and specific critiques of your drafts and specific suggested improvements.

I recall, in my very early thirties, starting a new position at a regional planning agency where one of my responsibilities was managing and then writing about our watershed planning projects. Soon after starting, I received one of my drafts back from the executive director and immediately noticed a big red "NG!" on the right side of a page. Clearly, it meant "Nice Going!" Wrong—on talking with the director, I learned he meant "No Good."

That was a minor setback because I quickly understood his concern and addressed it in the draft. He and I had many editing discussions over the years, I grew as a writer, led the writing of several major documents, and the agency's stakeholders were well served.

Expect a Wide Variety of Comments

As indicated at the beginning of this editing section, editors can choose to comment on many aspects of your writing. Furthermore, some selected aspects of editing are objective, such as spelling and grammar. Other topics, such as tone and relevance, are mostly subjective. The combined effect of the preceding means that if two or more editors review your draft, you are likely to receive and need to resolve a wide variety of comments.

Consider an example. I submitted a draft paper for possible presentation at a national engineering conference and then publication in the conference proceedings. Four anonymous editors reviewed the draft, and I received their written comments.

One reviewer said, "I found myself being bored by the papers I was receiving. Then I read this paper! I sincerely thank the author for writing such an interesting and insightful paper." Another editor stated, "While this will likely be an interesting presentation, it's not that insightful of a paper. The writing appears hastily constructed and at times very casual and conversational." Same draft, two very different reviews with one editor finding the draft insightful and the other editor finding the draft not insightful.

In summary, the four anonymous editors provided many comments, including some favoring acceptance of the draft and others rejecting it. I sorted through all the comments and used them to make many changes. The conference organizers accepted the significantly improved paper for presentation and publication.

If you proactively venture into the writing world, you will have similar experiences. As important as your ego may be, soothing it should not be your goal. Instead, view editors' comments—favorable and unfavorable—as resources for your use in producing documents that attract, inform, and influence readers. That's what communicative engineers do.

3.5 TIPS ABOUT SPECIFIC FORMS OF WRITING

3.5.1 Emails

I start with emails because, in the work world, they are the most common form of writing. Refer to the CSU Writing Guides [44] for comprehensive and detailed email writing advice, with helpful explanations.

Listed here, based on the CSU suggestions and my experience, are what I consider some of the most important email writing advice:

- Compose an informative and compelling subject line, in the spirit of Section 3.4.20. Recognize that your intended recipient(s) could easily receive dozens of competing emails every workday.
- State your purpose at the beginning of your message, then elaborate, and if you wish to share supportive detailed data or information, consider one or more attachments. In a sense, use a variation on the 3-part document structure described in Section 3.4.4.
- Conclude your email with a clear indication of what you hope will happen next. For example, you might write, "I hope this information is useful, please let me know if you have questions, suggestions, or concerns." Or you may make a specific request such as, "Please sign and return the attached agreement so we can promptly begin the project." As suggested by the two example quotes, consider adopting a respectful, helpful, and direct tone.
- Create and usually use a complete signature block at the end of the email. Consider including your name, title, organization, street address, telephone number(s), email address, and website. Enable the recipient(s) of your email to immediately and easily contact you or learn more about you and your organization.
- Avoid trying to use email or other social media to resolve interpersonal conflicts. The digital record of your comments will be in cyberspace forever. Days to years from now, your opponents or supporters may intentionally or

accidentally retrieve your message and quote it out of context to your detriment. While admittedly more difficult than writing and sending an email, a face-to-face (F2F) conversation is preferable because it includes voice tone and body language both of which enhance communication. If you insist on using social media to respond to a troublesome email or other message, draft it, "sleep on it," review it at least one more time, and then drop the idea or hit "send."

3.5.2 Memoranda

In contrast with emails, typically used to communicate both within and between organizations, memoranda are primarily for internal communication. These documents tend to focus on one topic and be short and to the point.

See Appendix F, Section F.2 for an example memorandum, in this case used to transmit the agenda for an upcoming meeting. Memoranda typically begin with a heading that includes the date, to, from, and subject. They often have attachments.

The suggestions I offered for writing emails generally apply to memoranda. That is, compose an informative and compelling subject line; state your purpose early on; use one or more attachments to provide supportive data or information; adopt a respectful, helpful, and direct tone; and conclude in an action-oriented manner. As with emails, CSU provides additional memoranda-writing advice [45].

3.5.3 Meeting Agendas

Let's define a meeting as three or more people discussing, F2F or virtually, professional or related topics. Meetings are common in the engineering world, beginning during engineering education. Because they require a major time investment, meetings should be carefully planned, conducted, and followed up. That process starts with a well-written agenda.

Absent an effective agenda—and a capable chairperson to orchestrate the event—meetings can be unproductive, a waste of valuable time, and frustrating to participants. As observed by comedian Milton Berle, "A committee is a group that keep minutes and loses hours."

Refer to Appendix F, Section F.2 of this book for a sample agenda. You, as a student or practitioner, will eventually be asked or volunteer to lead a group and, therefore, prepare agendas—this is inevitable. Therefore, note the following features of the sample agenda [46]:

1. Before even getting to the agenda, the "To" part of the transmittal memorandum lists all the individuals who are expected to attend the meeting. Invitees value this because they may have other matters to discuss with some of the attendees and can do so before or after the meeting.

2. The memorandum states both the date and day of the meeting and the expected start and ending time, all of which increase the likelihood that invitees will schedule the time slot accurately.
3. The meeting will begin with the topic "Good news." This is a personal preference—I have found that usually, one or more participants have relevant good news to share, which creates a positive atmosphere. No need to have "Bad news" on the agenda—that takes care of itself.
4. The "Additional agenda items" entry provides an opportunity for anyone to raise an issue they think is important.
5. The agenda is supported by three attachments (not shown) to encourage invitees to use in preparing for the meeting.
6. The agenda uses active verbs, such as discuss, select, follow-up, and decide, to encourage an informative and productive meeting.
7. The names of some invitees appear in the agenda to encourage them to effectively introduce and help resolve various issues. When we see our name in print, we are even more likely to be prepared.

Having mentioned minutes, make sure you document your meeting within a few days after the event. Recall the story of an undocumented meeting in Section 1.1.3. Because a consulting firm failed to document a client-consulting firm meeting, the firm incurred monetary and relationship costs. Undocumented meetings do not, for practical purposes, work well. Participants will remember each undocumented meeting in different and conflicting ways [47].

3.5.4 Letters

We typically use letters for communication from a person at one organization to a person at another organization. They serve a wide variety of purposes, such as seeking employment or offering it, accepting or refusing a job offer, ordering something or acknowledging its receipt, placing a complaint or responding to one, asking for assistance or receiving it, and transmitting a document or object.

Letters are more formal than emails and memoranda—we usually refer to the addressee with a Ms., Mr., or similar title. However, like emails and memoranda, the most effective letters have an inviting regarding (RE) description, state purpose, provide supporting data and information, and suggest action.

CSU Writing Guides provide broad and deep letter-writing advice [48]. Refer to Appendix F, Section F.3 of this book for a sample letter.

3.5.5 Letters to the Editor

Assume you just read an article, in a magazine or newspaper, about engineering or some other topic that motivates you to offer your pro or con views. Then consider sending, via email, a letter to the editor with the hope for publication in a near-future issue of the magazine or newspaper. Whether you take a pro or con position, your letter could help to encourage further productive discussion among the readership. You will also benefit because, as previously discussed in Sections 1.3.3 and 3.3, writing encourages broader and deeper thinking and learning about a topic.

Appendix F, Section F.4, provides an example letter to the editor. It illustrates key features of successful letters to the editor. More explicitly, letters selected for publication are:

- Short—typically under 200 words.
- Focused—references, very early, specific content, which recently appeared in the magazine or newspaper and which motivated writing to the editor. Others determine the subject, not the letter writer.
- Without sources—the content is largely the writer's opinion.
- Helpful—presumably the editor accepts the letter for publication because it includes new and credible content readers will value.

If your letter is accepted for publication, the editor may edit it slightly to conform with the writing style of the magazine or newspaper, shorten it, or clarify some content. The letters usually have a minimal signature, typically the writer's name and city.

3.5.6 Opinion Articles

These articles regularly appear in newspapers and other publications, like NSPE's *PE-The Magazine for Professional Engineers*. They are often called op-eds because, when published in newspapers, they appear on the page opposite the editorial page.

Opinion articles, as indicated by the title, are similar to letters to the editor in that they provide another way for you to share, via widely published writing, your views. Opinion pieces differ from letters to the editor in three major ways. You, the author, get to pick the topic; the allowable word count is much larger, which means that you can more fully address the topic; and you will be more fully described, sometimes including your contact information. The third point is important because writing opinion and other articles, which include information about you, gradually expands your network and, therefore, individuals with whom you may seek assistance or collaboration.

Refer to Appendix F, Section F.5, for an example of an opinion article published in NSPE's magazine. Note that the article:

- Describes my accidental discovery of a new-to-me subject (brain basics), indicates how brain basics can enable engineers be more effective, and urges engineers to consider learning more about their brains and how that knowledge could benefit them.

- Mentions one source in the text but, as is the practice, does not include a list of cited sources.
- Consists of just over 800 words—typical in that opinion, articles are roughly five to ten times the length of letters to the editor.

The example opinion piece, along with my other published writings, usually begins with me having and developing an idea. I prepare a proposal for an article, share the proposal with the editor of a magazine or other publication, often find interest, and work with the editor or staff to write, edit, and publish the article.

3.5.7 Peer-Reviewed Presentations and Papers

At some point in your career, you may want to present a peer-reviewed paper at an engineering conference and/or publish a peer-reviewed paper in an engineering journal. Members of a discipline generally consider peer-reviewed presentations and papers as the most credible sources in a field of study. At that point, in your personal and professional development, you may believe you have something to share with other engineers, such as how you and others solved a challenging problem, the findings of your research, or ideas on ways to improve engineering education and practice.

If you are an engineering student now, presenting or publishing a peer-reviewed paper may seem unlikely. Had I learned about peer-reviewed papers, when I was a student, I probably would have thought the topic to be irrelevant. However, as I proceeded in my career and had many different experiences, I increasingly wanted to share what I had learned and advocated, and, therefore, I have now presented or published many peer-reviewed and other papers. My presentations and publications enabled me to "road test" my content and views, experiment with various forms of speaking and writing, provide value to audience members and readers, meet some of the leaders of my engineering discipline, and travel to many interesting places.

Back to you. Peers, typically experts in relevant topics who volunteer to encourage quality presentations and publications, will review your draft submittal. Their collective reviews will help determine acceptance or rejection of your proposal to present or publish.

The peer review process typically includes meeting many requirements. A few examples: due dates, abstract content and length, maximum length of the paper, fonts, person (e.g., third person), citing of sources, and captioning of figures and tables. Make sure you understand and respect the review process requirements to avoid having your paper rejected because of failure to meet them.

Assuming you satisfy review requirements, have solid content, and write well, a peer reviewer will receive your proposal. Consider the reviewers to be conscientious editors, as discussed in Section 3.4.23. As also noted in that section, expect a wide variety of comments, some critical of parts of your submission. Objectively consider each comment, learn from it, revise the paper as appropriate, and resubmit if requested.

If your paper is approved for presentation at a conference, and then maybe publication in the conference proceedings, begin to think about how you will speak at the conference. Chapters 4 and 5 of this book offer, respectively, many speaking and visualization ideas.

3.5.8 Resumes

When I started writing about this topic, a quick Google search found over 60 books devoted to resume writing, suggesting that this is an important and complex topic. Essentially, all engineers, typically beginning as students and then throughout their careers, create resumes. They do this for a variety of purposes, send them to a variety of reviewers, and generate a variety of positive and negative results. Given the topic's complexity, I wrote this section to help you get started on writing your resume or to improve the current version. My "getting started" suggestions, based on my experience and CSU advice [49] follow.

1. Consider assembling and maintaining a master resume that you can repeatedly use throughout your career to create resumes that serve specific situations. Your master resume could include the following sections: Education, Employment History, Professional Registration (e.g., PE) and/or Certification, Professional Society Service, Public Service, Recognition/Awards, Areas of Expertise, Experience (e.g., projects you worked on and your contributions), Professional Memberships, and Presentations and Publications. Keep the master resume current so that it is always ready for use in creating a version for a specific use.
2. Construct your resume, or most portions of it, using a chronological format. For example, list your education in reverse chronological order. This format is common and emphasizes your most recent activities.
3. Use a functional format for portions of your resume, that is, describe your roles and what you and others accomplished. For example: Served as president of the NSPE student chapter and led a recruitment drive that produced a 30% increase in membership. Don't just list titles, like President of the NSPE student chapter. Work that in, as shown by the example, but emphasize what you did.
4. Note the active verbs "served" and "led" in the previous example. Recall the vigorous verb discussion in Section 3.4.7. Whenever you describe what you did or the recognition you received, start with or otherwise use an active verb and be specific. For example:
   - Assisted Professor Smith with her robot research project by programming its walking mode

- Drafted a proposal that helped our consulting firm obtain a contract to design a manufacturing line
- Applied the five-why analysis to determine the root cause of a problem in a chemical process

5. Before tailoring your master resume for a particular situation, define the audience and the purpose.

During discussions of resumes, someone may mention Curriculum Vitae (CV), which is Latin for the course of your academic life. Academia uses this type of resume when someone, such as you, is pursuing a teaching and research position or applying to graduate schools. CVs tend to be larger than resumes because they include long and detailed lists of courses taught, research conducted, papers published, and graduate students advised [50]. The advice offered for resumes is generally applicable to CVs.

3.6 KEY POINTS

- Writing is an essential part of formally studying and later practicing engineering.
- Writing uses mostly one channel (verbal, that is, words) and is one-directional (from writer to reader), while, in sharp contrast, speaking uses three channels (verbal, voice, visuals) and is two-directional (from speaker to listener and vice versa).
- The communicative engineer is adept at writing and speaking and knows when and how to use each of these very different modes.
- While students and practicing engineers could view learning how to write as a life-long endeavor, they could also view writing as a means of life-long learning—of broadening and deepening knowledge of whatever topics interest them.
- The chapter includes about two dozen immediately applicable items of writing advice for students and practitioners to use, as they see fit, in improving their writing or helping others improve theirs. It also offers tips for specific forms of writing.

Writing is a tool that enables people in every discipline
to wrestle with facts and ideas.
It's a physical activity, unlike reading.
Writing requires us to operate some kind of mechanism—
pencil, pen, typewriter, word processor—
for getting our thoughts on paper.

—William Zinsser, writer, editor, teacher

REFERENCES

1. Decker, B. (1992). *You've Got to Be Believed to be Heard.* New York: St. Martin's Press.
2. Walesh, S.G. (2012). *Engineering Your Future: The Professional Practice of Engineering.* Chapter 3, Communicating to make things happen. (pp. 81–82). Hoboken, NJ: Wiley. Figure 3.2 is from this book and the concept is used with permission from ASCE.
3. Lewis, M. (2017). *The Undoing Project: A Friendship That Changed Our Minds.* New York: W. W. Norton & Company.
4. Zinsser, W. (1988). *Writing to Learn.* New York: Harper Resource.
5. Strunk, W. and White, E.B. (1919). *The Elements of Style-Fourth Edition.* (pp. 18–19). New York: Allyn and Bacon.
6. University of Chicago. (2017). *The Chicago Manual of Style.* Chicago: University of Chicago Press.
7. Colorado State University. (2023). Writing guides. *Writing@CSU.* https://writing.colostate.edu/guides/ (accessed 26 June 2023).
8. Johnson, P. (2009). Walking our way out of recession. *FORBES.* (September 21).
9. Walesh, S.G. (2017). *Introduction to Creativity and Innovation for Engineers.* Chapter 2, The brain: a primer. (pp. 37–42). Hoboken, NJ: Pearson Education.
10. Walesh, S.G. (2004). *Managing and Leading: 52 Lessons Learned for Engineers.* Lesson 32, The power of our subconscious, Reston VA: ASCE Press.
11. Walesh, S.G. (2017). *Introduction to Creativity and Innovation for Engineers.* Chapter 4, Basic whole-brain methods. (pp. 123–127). Hoboken, NJ: Pearson Education.
12. Nagle, J.G. (1998). Seven Habits of Effective Communications. *Today's Engineer.* Summer.
13. Walesh, S.G. (2004). *Managing and Leading: 52 Lessons Learned for Engineers.* Lesson 21, Start writing on day 1. Reston VA: ASCE Press.
14. Vesilind, P.A. (2007). *Public Speaking and Technical Writing Skills for Engineering Students.* Woodsville, NH: Lakeshore Press.
15. O'Connor, P.T. (1999). Words Fail Me: What Everyone Who Writes Should Know About Writing. San Diego, CA: Harcourt.
16. Greenman, R. (2005). Words that Make a Difference and How to Use Them in a Masterly Way. (pp. 395–396). Delray Beach, FL: Levenger Press.
17. Merriam-Webster. (2023). Point of view: it's personal – first, second, and third person explained. https://www.merriam-webster.com/words-at-play/point-of-view-first-second-third-person-difference (accessed 29 March 2023).
18. Osmond, C. (2023). Writing in first, second, and third person – ultimate guide. https://grammarist.com/grammar/first-second-and-third-person/ (accessed 29 March 2023).
19. Walesh, S.G. (2012). *Engineering Your Future: The Professional Practice of Engineering.* (pp. 88–89). Hoboken, NJ: Wiley.
20. Merriam-Webster. (2023). https://www.merriam-webster.com/dictionary/metaphors (accessed 30 August 2023).
21. Rico, G. (2000). *Writing the Natural Way: Using Right-Brain Techniques to Release Your Expressive Powers.* New York: Jeremy P. Tarcher/Putnam.
22. Merriam-Webster. (2023). https://www.merriam-webster.com/dictionary/punctuation (accessed 26 May 2023).
23. Truss, L. (2009). *Eats Shoots & Leaves.* New York: Harper Collins.

24. Walesh, S.G. (2018). Writing: how to engage and convince your readers. ASCE. webinar presented 11 October 2018.
25. Clark, B. and Crossland, R. (2002). *The Leader's Voice*. New York: Select Books.
26. Walesh, S.G. (2012). *Engineering Your Future: The Professional Practice of Engineering*. Chapter 11, Legal framework (pp. 341–342). Hoboken, NJ: Wiley.
27. Hayden, Jr., W.M. (1987). Quality by Design Newsletter, May.
28. Bachner, J.B. (2007). Writing well to avoid risks. *CE News*. September.
29. Merriam-Webster Dictionary. (2023). https://www.merriam-webster.com/dictionary/euphemism (accessed 2 February 2023).
30. Loeffelbein, B. (1992). Euphemisms at work. *The Rotarian*. February.
31. Merriam-Webster. (2023). https://www.merriam-webster.com/dictionary/specificity (accessed 10 April 2023).
32. Berthouex, P.M. (1996). Honing the writing skills of engineers. *Journal of Professional Issues in Engineering Education and Practice*. American Society of Civil Engineers, July.
33. Walesh, S.G. (2004). Force or finesse? Forum section. *Leadership and Management in Engineering*. ASCE. October.
34. Walesh, S.G. (2004). *Managing and Leading: 52 Lessons Learned for Engineers*. Lesson 22, P^5: Preparing, presenting, and publishing professional papers. Reston VA: ASCE Press.
35. Walesh, S.G. (2021). *Engineering's Public-Protection Predicament: Reform Education and Licensure for a Safer Society*. Valparaiso, IN: Hannah Press.
36. Grandin, T. (2022). *Visual Thinking: The Hidden Gifts of People Who Think in Pictures, Patterns, and Abstractions*. New York: Riverhead Books.
37. King, S. (2000). *On Writing: A Memoir of the Craft*. New York: Scribner.
38. MarketingSherpa, Inc. (2003). High-impact email writing part 1: useful lists of short words, strong verbs, and blah words. e-newsletter, August 13.
39. Walesh, S.G. (2007). Writing – less is more – part 1. *Indiana Professional Engineer*, March/April.
40. Alexander, D. and Rivett, A. (1998). The tin ear. *Journal of Management in Consulting*. November.
41. University of North Carolina. (2023). Reading aloud. The Writing Center. https://writingcenter.unc.edu/tips-and-tools/reading-aloud/ (accessed 5 April 2023).
42. Wikipedia. (2023). Editing. https://en.wikipedia.org/wiki/Editing (accessed 31 January 2023).
43. Sante, L. (2012). Finding the editor within. Word Craft column. *Wall Street Journal*. March 10-11.
44. Kiefer, K., Kowalski, D. and Bennett, A. (2022). Email. *Writing@CSU*. Colorado State University. https://writing.colostate.edu/guides/guide.cfm?guideid=44 (accessed 26 June 2023).
45. Connor, P. (2009). Business memos. *Writing@CSU*. Colorado State University. https://writing.colostate.edu/guides/guide.cfm?guideid=73 (accessed 26 June 2023).
46. Walesh, S.G. (2012). *Engineering Your Future: The Professional Practice of Engineering*. Chapter 4, Developing relationships (pp. 135–140). Hoboken, NJ: Wiley.
47. Walesh, S.G. (2004). *Managing and Leading: 52 Lessons Learned for Engineers*. Lesson 39, An "unhidden" agenda. Reston, VA: ASCE Press.

48. Connor, P. (2009). Letters. *Writing@CSU.* Colorado State University. https://writing.colostate.edu/guides/guide.cfm?guideid=71 (accessed 9 July 2023).
49. Barton, S.M. (2003). Resumés. *Writing@CSU.* Colorado State University. https://writing.colostate.edu/search/index.cfm?q=resumes (accessed 9 July 2023).
50. Rallo, R. (2005). Curriculum Vitae. *Writing@CSU.* Colorado State University. https://writing.colostate.edu/guides/guide.cfm?guideid=62 (accessed 26 June 2023).

EXERCISES

3.1 Apply mind mapping: Think of a subject of interest you want to learn more about, and eventually summarize what you learn in a written report. Use mind mapping to generate ideas, questions, sources, topics—anything even remotely related to the subject. Include whatever "pops into your head," go for quantity, not quality, and resist the temptation to evaluate whatever you add to the evolving mind map. Work on the map for, say 15 minutes, set it aside, do something else, and return to it hours later. Repeat that take-a-break process as long as it yields more content. Do not be concerned about the neatness of your mind map. Submit the mind map and your written observations about the positive and/or negative aspects of the process.

3.2 Apply artificial intelligence: Use ChatGPT, or a similar AI system, to repeat Exercise 3.1. Compare the results, indicate which approach—mind mapping or AI—provided the best results from the perspective of finding content about your topic of interest, and explain why.

3.3 Research artificial intelligence: Do a more modest version of Exercise 1.6 and focus on some aspects of AI. This effort will provide more research and writing practice and help you keep up with, and maybe contribute to, this evolving new technology.

3.4 Evaluating writing advice: Skim Section 3.4 and decide what the most valuable writing advice is for you right now. Indicate what you selected and why.

3.5 Create a vigorous verb: Select a noun connected to your engineering or related discipline and readily recognized by its members. Write a sentence that uses the noun as a noun. Then write an active voice sentence that uses the noun, or a variation on the noun, as a verb.

3.6 Writing to different audiences: You just learned, in one of your engineering classes, about liquid flow through a venturi, including why the pressure in the narrowest section of a venturi is lower than before or after that section. An engineer friend missed that class and asked you what the instructor presented. On learning that one topic was venturis and that the pressure is lowest in their narrowest parts, your friend expresses doubt—doesn't seem logical because the liquid is being "compressed" as it approaches the narrowest section. Write a few sentences for your engineer friend, including the use of equations, to explain why it is logical.

Now a nontechnical friend asks about that class as in "what did you learn today?" You briefly describe a venturi and, when you mention that the pressure is lowest in the narrowest part, your friend expresses doubt. Write a few sentences for your nontechnical friend, without using equations or formulas, to explain why it is logical. (Note: The plural of venturi is venturi or venturis.)

3.7 Writing improvements you would make: Recall and find something you recently wrote. Some examples include a research paper you wrote for a class, a memorandum you wrote to your supervisor, a laboratory report, a letter transmitting your resume to a potential employer, or a capstone course report. Based on the writing advice offered in this chapter, describe up to three things you would change.

3.8 Use a metaphor to strengthen a sentence: Skim some text—text you or someone else created. Find a sentence that does not include a metaphor. Revise the sentence so that it includes a metaphor that makes the sentence stronger. For example, I wrote this sentence in one of my published books: "I especially wanted to reach those who are dissatisfied with the state of U.S. engineering and, more importantly, are open to changing licensure law and reforming education for licensure." I could have made the sentence stronger by including two metaphors, as follows: I especially wanted to *grab readers* who are dissatisfied with the state of U.S. engineering and, more importantly, would *jump at the chance* to change licensure law and reform education for licensure.

3.9 Use a simile to strengthen a sentence: Skim some text—text you or someone else created. Find a sentence that does not include a metaphor or simile. Revise the sentence so that it includes a simile that makes the sentence stronger. Share your thoughts about metaphors and similes. Which one is more effective, and which is more difficult to create?

3.10 Read your writing out loud: Select several pages from one of your final documents in Exercise 3.7. Now read it out loud and, if you make any corrections or changes, note how many changes you made and what they were. Summarize the number and nature of the changes and share your initial view—positive or negative—of the value of reading your writing out loud.

CHAPTER 4

SPEAKING

> If I am to speak ten minutes,
> I will need a week's preparation.
> If 15 minutes, three days.
> If half an hour, two days.
> If an hour, I am ready now.
>
> —*Woodrow Wilson, 28th U.S. President*

After studying this chapter, you will be able to:

- Identify the beneficiaries of effective speaking
- Provide examples of situations in which speech is the preferred communication mode
- Describe how you can use speech preparation to learn
- Discuss the three parts of an effective speaking project
- Use one or more of the speaking suggestions described in this chapter to assist someone who is committed to becoming a more effective speaker

4.1 SPEAKING'S ROLE IN ENGINEERING

4.1.1 Speaking Defined

We turn now to speaking, which is generally regarded as the most challenging of the five modes of communication—asking, listening, writing, speaking, and using visuals—discussed in this book. I say "most challenging" because about three-fourths of the population experience glossophobia, the fear of public speaking. Possible glossophobia symptoms include panic, elevated blood pressure, anxiety, nausea, perspiration, dry mouth, and stiffening of

The Communicative Engineer: How to Ask, Listen, Write, Speak, and Use Visuals, First Edition.
Stuart G. Walesh. © 2024 John Wiley & Sons, Inc. Published 2024 by John Wiley & Sons, Inc.

upper back muscles [1]. More about glossophobia and how to deal with it later in this chapter.

Speaking means "conveying information or expressing one's thoughts and feelings in spoken language" [2]. For the purpose of this chapter, we qualify that definition by limiting it to informal or formal presentations to an audience in business, government, academic, or similar settings. Thus, we distinguish speaking from the one-on-one questioning and answering explored in Chapter 2.

4.1.2 Required and Optional Speaking

Most engineering students will be required to do some speaking, and all will have opportunities to speak when not required. Some examples of both:

- Completing an assignment in an engineering or other course and summarizing the results in a presentation to the class
- Taking an independent study course or working on a master's degree, conducting research, preparing a report, and sharing what was learned by speaking to faculty and students
- Volunteering, as a senior, to be the presenter for a team in a capstone course

I vividly recall my first speaking assignment in college—it was in a one-credit speech course required for sophomore engineering students. My first speaking assignment: Explain how a nuclear power plant works—I had no idea. However, after some research (no internet), I outlined a presentation, prepared a diagram of the nuclear power process on a whiteboard, practiced, fought nervousness, presented, and survived. My second presentation was a little bit easier, and then I was on my way to continuously improve my speaking knowledge, skills, and attitudes (KSA).

Once engineers enter practice or take a position in academia, speaking opportunities abound, such as:

- Sharing lessons learned, on a just-completed project, with other staff at a noon-hour "brown bag" session
- Being one of several engineering firm members speaking, as a team, to leaders of a potential client for the purpose of winning a first project with them
- Giving a talk at the weekly meeting of a local service club about the implications of some new technology
- Speaking, at a national conference, about a creative solution to a challenging engineering problem

Your effective speaking can lead to win–win–win–win results. Consider each of the four "wins" [3].

1. **You** "win" in that you learn because of preparing, speaking, participating in a question and answer (Q&A) session, and following up after the event. You also "win" by becoming more confident every time you

speak, even when you are not as successful as hoped, because of what you learn from a setback.

2. Your **audience** "wins" because they learn based on your study, preparation, and delivery. In some cases, you will be respectfully and fondly remembered as a teacher and influencer.
3. Your **employer** "wins" because when you speak, you are temporarily your organization in the eyes and minds of audience members. Your effective presentation strengthens the organization's reputation.
4. The **profession, and public it serves,** "wins" if your presentation, or portions of it, are published in widely available forms such as conference proceedings, published papers and articles, and books.

A reminder: Recall the Chapter 3 discussion of the differences between writing and speaking (Section 3.2). More specifically, if you are in a situation where your primary goal is sharing data or information with one or more individuals, then writing is likely to be the preferred communication mode. In contrast, if you and others are advocating major change or a completely new innovative or creative approach, then rely mostly on speaking, at least early on.

4.1.3 Commit Now and Avoid Regret Later

Frankly, becoming a good to great speaker requires commitment and correspondingly consistent effort. While we may admire great golfers, chefs, artists, singers, and others, very few admirers will achieve what they admire. So it is with speaking and engineers—very few engineers will become great speakers.

I mention the preceding now, while you are an engineering student or young practitioner, because, if your goal or even inclination is to become a great speaker, now is the time to start. You have many years ahead of you and have this book in your hands while reading this speaking chapter. Do I say that because I believe this is the best-speaking chapter ever written? No, not necessarily—time will tell. But this is the best-speaking chapter I was able to write now for your possible use.

Avoid regret later. Consider committing to becoming a great speaker and then having many doors open to you.

4.2 LEARNING TO SPEAK AND SPEAKING TO LEARN

As explained in the previous chapter (Section 3.3), while we want to learn how to write, we soon realize that we can also do the reverse—use writing to learn about any topic of personal interest. The same applies to speaking.

You may be tempted to speak only about subjects in which you have broad and deep knowledge, because this approach minimizes preparation time and effort. This safe tactic may, in special circumstances, be appropriate. However, if your goal is continuous learning and personal growth, then seek speaking opportunities that are likely to lead you to new knowledge. Frankly, even when we agree to write or speak within our area of expertise, we will find knowledge gaps that require filling and, therefore, more study.

4.3 SPEAKING ADVICE: INTRODUCTION

The principles of effective communication described in Section 1.5, especially the first three—know audience, state purpose, and accommodate preferred ways of learning—should be reviewed before selecting and using this chapter's speaking advice. Then build on those principles using the advice.

Visuals such as photographs, tables, graphs, props, and various figures are essential for effective speaking. Chapter 5 describes the reason for that observation and shows how to prepare and use visuals. Visuals are so important that they warrant a chapter.

Think of speaking as a three-part process, as illustrated in Figure 4.1, consisting of preparation, presentation, and following-up. Each part requires thoughtful attention, with the first part demanding, by far, the most effort and immediate follow-up the least effort.

I heard a thought-provoking and effectively delivered sermon. Immediately after, during lunch, I happened to sit next to the pastor and asked if he would explain his sermon preparation process. He described working over a week on each sermon for a total preparation time of about 20 hours. That is 20 hours of preparation for 20 minutes of delivery giving a preparation-to-delivery ratio of 60. On the heavy side, but not beyond reason depending on how new the subject matter is to the speaker. My rule of thumb ratio is 20 for a new presentation, assuming I already have most of the content, am adding some new content, and am tailoring the entire presentation to the audience [3].

You say that's a lot of work for one presentation! I say, you bet! Based on my experience and my knowledge of accomplished speakers, I also say the following: That kind of speaking preparation attitude and effort will open doors

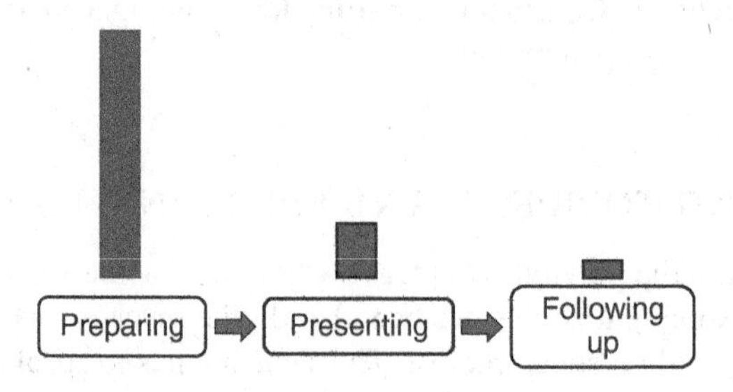

Figure 4.1 This three-part process, with preparation requiring the most effort, produces an effective presentation.

for you, put you in positions of influence you may never have envisioned, and take you around the globe.

Pragmatic speaking advice follows and is organized by the three parts of the speaking process. View the suggestions as a toolbox from which you select what you need to craft, deliver, and follow up on a win–win–win–win presentation.

4.4 SPEAKING ADVICE: PREPARING

4.4.1 Get on the Program

If you want to speak in situations beyond where you are required to speak, then you need to "get on the program," that is, find an audience. One way to "get on the program" is to be invited. If you want to become a proficient speaker, you will never make it waiting to be invited. Why? We begin as amateur speakers who are not likely to receive many speaking invitations.

Therefore, whether you are or become a practitioner in business, government, or academia, achieve your communication improvement objective by inviting yourself in ways such as these:

- Tell your "boss" and co-workers—or your professors and fellow students—about your interest in becoming a better speaker. Mention topics about which you are knowledgeable or want to learn more about.
- Inform business, government, and education leaders in your community about your speaking interest. For example, early in my career, I spoke to fifth graders about civil engineering because their teacher was a friend of mine.
- Share your speaking interest with professional society committees. You could start with the program chair of your society's local branch or section because they are often looking for speakers.
- Look for formal "calls" for papers if you want to speak at state, national, or international conferences. These notices typically appear on professional society websites or arrive via email six months or more before the event. A call typically announces the conference place, dates, theme, and principal topics. The notice invites abstracts or short descriptions of proposed papers and provides a schedule for submission and review of the abstract and, if accepted, for providing a complete manuscript. Some conferences publish the accepted and delivered papers in proceedings. Figure 4.2 illustrates the ideal way to respond to a call for papers. It also shows a common, less desirable, but feasible and acceptable approach, which assumes you have the commitment and knowledge needed to write a paper and prepare a presentation should your abstract be accepted. The principal difference is when the paper is drafted—early in the ideal approach and late in the common approach.

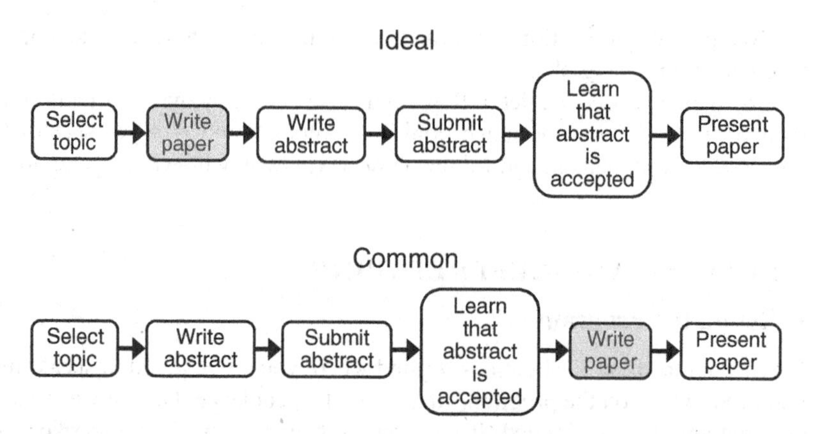

Figure 4.2 The ideal and common ways to respond to a call for papers, with the main difference being when the proposer completes the first draft of the paper.

Assume you are a practitioner who just learned that you will speak at a conference, or you are a student who will speak in class, on campus, or in your community. You are beginning to prepare your presentation or, if you already have a rough draft, are improving it. Then consider suggestions for the preparation part of your speaking project.

4.4.2 Tell–Tell–Tell

Either directly or in a subtle manner, structure your presentation so that you tell the audience what you are going to tell them, then tell them in the main part of the presentation, and, finally, you conclude by telling them what you told them [3]. The first "tell" could be, or include, the upfront, clearly stated purpose of your presentation.

The T^3 approach's repetition helps audience members know where you are going, recall where you have been and, when you have finished speaking, remember your principal message. Another reason for T^3 is that it reduces audience anxiety about what is on your mind and what you might ask them to do.

4.4.3 Generate Content

Review the three content-generation methods described in the writing chapter (Section 3.4.2) and consider using one. Those tools are applicable to preparing a speech. I prefer and often use mind mapping, the second method, for preparing a speech or when I want to generate ideas for other purposes. An individual or a group can use this highly visible and nonlinear tool. Keep track about what is happening in AI because, as explained in Section 3.4.2, AI has the potential to help you obtain content ideas.

4.4.4 Begin to Think about Visuals

As defined again, and then discussed in detail in Chapter 5, a visual is something appealing to the sense of sight that we use when speaking or writing to inform, convince, or otherwise engage others. Now is the time to start thinking about how visuals could help you achieve the purpose of your presentation. Chapter 5 will assist you with the details.

What do I mean by "start thinking" about visuals? Assume that you are a student preparing a speech describing the process you used to complete a one-credit independent study course. You have outlined what you are going to say. As you describe each of the many steps, you decide to use a flow chart that, using the animations feature in PowerPoint, introduces each step and its relationships to other steps. Chapter 5 offers suggestions for preparing that and other visuals.

4.4.5 Use Metaphors and Similes

Chapter 3 (Section 3.4.10) defines metaphors and similes, provides examples, and suggests using them in writing. These helpful figures of speech are even more effective when speaking because you can increase their impact with visuals.

For a metaphoric example, consider a presentation I made that included the interaction between our conscious and subconscious minds [4]. Neuroscientists discovered that the thinking of our conscious mind, which we are aware of, is a tiny fraction of our total thinking. The vast majority of our thinking occurs, without our knowing it, by our subconscious mind. I showed a cross-section view of a floating iceberg and noted that the tip is our conscious thinking and the huge submerged portion is our subconscious thinking.

Now for a simile example: You, a practicing engineer, while speaking to engineering students about engineering licensure, say that earning an engineering baccalaureate degree and not taking the fundamentals of engineering (FE) examination is like finally buying your dream car and not taking it for a ride. Simultaneously, you reinforce your words with colored images of one or more dream cars. Audience members will tend to remember one or more of the car images, the illogical action of earning one and then not driving it, and questionable practice of earning an engineering degree and not taking the FE examination.

Consider another simile example. The Milwaukee, WI, area was planning to use a deep tunnel bored in bedrock to temporarily store and convey combined sewage prior to treatment. The system is now in operation. Citizens were concerned with leakage of combined sewage from the tunnel into the surrounding aquifer. The tunnel was to be about 200 feet below the water table. Someone on our project team suggested using a submarine simile. That is, the tunnel would be like a submerged submarine in that if leakage occurred, it would be inward—not outward. The simile proved to be an effective means of communication.

I recognize that images can also be used in documents, but the number is typically limited, what we show is static, and color may not be feasible. In contrast, when you or I speak to an audience, we can easily use many colored images, some with animations, props, and body language to supplement what we say. Audience body language and verbal responses indicate how they receive our message. When speaking to an audience, we have the ability to create a communication-rich environment for our and their benefit.

4.4.6 Tell True Personal Stories

This is the same advice I offered in the writing chapter (Section 3.4.13), and for the same reason—true personal stories combine facts and feelings—speak to head and heart. They engage audience members, who then typically understand and often remember the stories' messages. When you speak, you tell your story face-to-face (F2F), and it is received on three channels. It will be even more effective than in a document.

Reminder: A story doesn't prove anything because it typically describes an isolated event or incident. The principal value of a true personal story and a visual is that you use them to explain to your audience, in a memorable cognitive and emotional manner, an idea or principle of potential value to them. The visual you use enables audience members to understand your message and recall it much later.

4.4.7 Provide a Handout

Briefly stop the presentation preparation process to ask yourself if a summary of your message—a takeaway—would serve your purpose and meet audience wants or needs. You could distribute the summary before, during, or after the presentation. I answer "yes" whenever I've asked that question and, therefore, always provide a handout or other takeaway.

A handout offers the following benefits [3]:

- Enables those who hear and appreciate, or at least are interested in, your message to "take it home" for possible study, application, or other use.
- Connects you to audience members, so that they or others can contact you after the presentation. This assumes that the handout includes information such as presentation title, date, event, and sponsor, plus your name, title, credentials, employer, email, telephone number, and website.
- Provides a backup in case the presentation system, such as the computer and/or projector, fails. I've had a computer fail when speaking to an audience and a lost internet connection when presenting a webinar, and was able to carry on with me and audience members using the handout. If you are using slides, number them for easy reference in emergencies.
- Reduces audience member note-taking or provides an effective place to write notes, such as the three-slides-per-page PowerPoint format.
- Causes you to think even more about the audience and how they might receive and follow up on your speech. That perspective may cause you to modify some of your content or means of presentation.

The handout could be as short as a one-page summary of your key points, preferably including some meaningful images. Or provide a complete set of slides that you used, preferably as a portable document format (PDF) file. If you will use props instead of, or in addition to, slides, then show some of the props on the one-page summary.

Organizers of state and national conferences typically request a slide set, written paper, or other materials prior to the event and then send them electronically to those who have registered for a conference. I've presented hundreds of webinars, and registrants almost always receive the slides as PDF files prior to the event or have the option of acquiring the slides after it.

When speaking F2F, and depending on the content, consider providing a handout at the end of the session. Reason: Having the handout at the session's beginning will distract some audience members. Therefore, you and they don't fully benefit from the valuable verbal and visual interaction inherent in a F2F setting.

4.4.8 Apply Advice Previously Offered for Writing

Chapter 3 provides 23 items of writing advice, some of which are directly applicable to preparing for speaking. For your convenience, a list of the applicable items, including their sections in Chapter 3, follows.

- Tell True Personal Stories (Section 3.4.13)
- Craft Informative Titles for Figures and Tables (Section 3.4.14)
- Avoid Liability (Section 3.4.15)
- Give Credit (Section 3.4.16)
- Minimize Euphemisms (Section 3.4.17)
- Strive for Specificity (Section 3.4.18)
- Use History to Support Your Purpose (Section 3.4.19)
- Title to Attract Attention (Section 3.4.20)

Consider looking at this advice as you prepare for your next speaking engagement.

4.4.9 Note This About Notes

Begin to think about how you are going to deliver the speech. According to Jack Valenti, who was the chief speechwriter for President Johnson and authored a now somewhat dated (published in 2002) but still useful book [5], you can choose from three basic delivery methods.

The first is to write out your speech until you are satisfied, and then **read it to the audience**. He would offer this advice: Don't do this unless you absolutely have to, as in a formal hearing. Lack of eye contact alone will separate you and your audience. Some participants may think that someone else prepared the speech and, therefore, you are just a messenger. The only delivery worse than reading your speech to the audience is showing all-text slides and reading them to the audience.

The second option is to write your speech, memorize it, and **recite it to your audience**. Typically, they will immediately notice the memorization, which will disconnect you from them. Valenti rarely uses this approach and only under conditions in which he is "totally and completely prepared," including the content and cadence, sentence length, pace, pitch, pauses, and body language.

The third option, and, in my view, the one chosen by the vast majority of engineers, is to **use notes skillfully**. The principal advantage of notes is that you can frequently look at them and speak directly to the audience. You might begin, as in the preceding two options, by writing your speech and then creating notes that capture essential content. Sometimes a written version is required, as when conference organizers accept your proposed presentation and will publish the written versions of all the conference presentations.

As you begin to practice your speech (see the next topic), make sure that you use your notes as a guide while looking often at the audience. Applying the "eye-five" method is one way to encourage you to look frequently at and around the entire audience. By "eye-five," I mean gradually gazing around the room, stopping occasionally to make contact for five seconds with various individuals, and moving on.

4.4.10 Practice Out Loud

Chapter 1 includes Practice Perfect Practice as one of the six principles of effective communication. Of the five communication modes presented in this book, this practice principle is most applicable to speaking. Briefly stated, before you speak in an F2F or virtual setting, practice out loud several times.

"Out loud" means exactly that—stand up, if that is how you will finally speak, imagine the audience in front of you, speak, gesture, advance your slides, display your props, and manipulate whiteboards. If you have never used out-loud practice, it will seem awkward the first few times. That discomfort will disappear when you experience the benefits over a series of presentations.

Ask a colleague, friend, or special someone to watch and listen to some of your practice presentations and then suggest improvements. If that approach is not feasible, consider making audio-video recordings of your practice runs or speaking in front of a mirror while recording your voice.

Consider another or supplemental potential approach to out-loud practice. "PitchVantage (PV), an e-learning platform, "uses artificial intelligence (AI) to deliver instant feedback on presentation skills. It works with companies and universities to develop great presenters" [6]. PV evaluates aspects of a speaker's delivery such as engagement, eye contact, pace variability, pauses, pitch variability, verbal detractors, and volume variability [7]. Your academic, business, or government organization could obtain PV, or a similar AI product, and make it widely available to interested personnel.

Having mentioned audio, appreciate that we may not know how our voice sounds to others. It may be credible and comforting, untrustworthy and annoying, or somewhere in between. When we talk, we hear our voice mostly after conduction through the bones in our head. Others hear our voice after transmission through air [8]. At minimum, know how your voice sounds to others by recording it or asking others. If needed, take corrective action.

Out-loud practice provides the following three benefits [3, 9]:

Set presentation time: If, as part of your preparation to speak, you read your presentation or speak it to yourself, you will greatly underestimate delivery time. Find out how much time organizers allot for your presentation, including a hoped-for productive Q&A session. Then use practice to make changes to your presentation so that you stay within that time limit. Respecting your scheduled time means that you respect the audience, and if there are other speakers, you respect them. Not honoring your allotted time by exceeding it sets off a domino effect that denies other speakers their full-time slots, and audience members fail to hear all speakers and/or all of their intended content.

Discover distractions and fix them: When speaking, we have many concerns with the most important being presenting our content and noting its reception. Accordingly, we may not recognize things we say and do that distract some audience members from giving us their full attention. Some examples: Frequently saying “ah,” “um,” “currently,” and “you know;” monotone speaking; focusing on just one part of the audience; rattling our keys; taking our glasses off and on; and looking at the screen instead of the audience. You can easily eliminate distractions—if you know about them. Out-loud practice, heard and viewed by others or, if recorded, observed by you, will reveal distractions, enable you to eliminate them, and greatly improve your speaking KSA.

Eliminate the rough edges: If you practice your speech out loud, only you, plus someone else if you have an observer, will hear or see your first presentation. Even though you have a definitive purpose, know your subject, and think you have the words and visuals to present it, some words won’t convey your intent, and, at other times, you will be at a loss for words or mispronounce some. Informed by the first practice run, the second time through will produce a greatly improved result, and each subsequent practice provides additional improvement.

Fortunately, for you and others, that first rough presentation will not see the light of day. Two or more practice runs will prepare you for prime time—for the intended audience. Minus out-loud practice, the audience is very likely to get that first rough version, in which case three entities lose—you, your organization, and the audience.

Refer to Appendix G for examples of the kinds of avoidable things that tend to happen in the absence of out-loud practice, which I assume did not occur or, if it did, was poorly conducted. Practice could have detected and, therefore, prevented essentially all of the many unfortunate speaking flaws.

Having stated the benefits of out-loud practice, consider some thoughts about how to practice. Assume your presentation has these six major sections: Problem description, solution 1, solution 2, solution 3, recommendations, and wrap-up. Segment your practices, that is, go through the first section, note and make needed improvements, go through it again, and record the elapsed time. Repeat for the other five sections. Now decide which sections need additional work, prioritize them, make changes, and practice each one again. This segmented approach exemplifies the perfect practice or deliberate practice described in Section 1.5.6, that is, fixing the weakest links first. Finally, practice the entire presentation out loud.

You may be thinking, that's a great but seemingly unreasonable effort—I wish I was one of those gifted speakers. None of the great speakers I've known, heard, or learned about were gifted. Consider the following related information:

- Winston Churchill, the former British Prime Minister, whose speaking ability helped lead his country through WWII, practiced an hour for every minute of his speeches [10].
- Former Apple Chief Executive Officer (CEO) Steve Jobs offered conversational and passionate introductions of new products. His team prepared "like mad to make sure it [looked] easy" [11].
- Ralph Waldo Emerson, schoolmaster, minister, lecturer, and writer, succinctly and bluntly stated, "All great speakers were bad speakers first."

4.4.11 Arrange On-Site Logistics

Physical arrangements will influence the success of your presentation. Therefore, well before your speech, discuss logistics with the person who invited you or someone else in charge of the event. Consider the following topics, based on my experience and the suggestions of the previously mentioned accomplished speaker, Jack Valenti [5]:

- Start and finish times include how long before the start time you can have access to the room to confirm logistics and how long after the finish time you have to remove your things from the room and converse with some audience members.
- Confirm that there will be a Q&A period.
- Computer you will use, if needed, such as yours or theirs.
- Remote control for computer.
- Projector, if needed.
- Podium, table, or both, depending on whether or not you are using props or have other needs. If you will be using your or someone else's computer

for any part of your presentation, make sure it will be placed (on the podium or a table) between you and the audience. When you show an image on the screen and talk about it, you want to see the image on the computer as you speak directly to the audience—not to the screen. If you talk to the screen, you show your back to the audience.

- Provision by you of a PDF file of the slide set for uses such as sending to participants prior to the presentation, copying and distributing to participants at the beginning or end of the presentation, or sending to interested participants after the presentation.
- Distribution by you of a special, perhaps summary, handout that you bring to the event and make available before or after you speak.
- Sound system, if needed, and type of microphone, such as fixed, handheld, and lavalier. The lavalier is best because it allows you to move around as you speak and during the Q&A period.
- Room lighting system—sometimes the image projected on the screen is "washed out" by nearby bright overhead lights.
- Flip chart on tripod.
- Extension cord.
- Seating arrangements such as chairs at tables, individual chair-desk sets, or chairs arranged classroom style or in a semi-circle.
- Availability of technical help immediately before and during your presentation.
- Find out who will introduce you and what you can provide to help them—such as the next item.
- Your very brief bio—several sentences—for possible use by the host or moderator when advertising the presentation and/or introducing you immediately before you speak. Audiences seem to like to know "who is this person?" Whenever I speak, I would like the audience to know that I am a PE, and have worked in academia, government, and business, and am an independent consultant, teacher, and author.
- Determine if there will be multiple speakers and, if so, who will speak before and after you and what will they speak about. If appropriate, while you speak, you may be able to reinforce a point made by the previous speaker or help transition to the next speaker.
- Evaluation by attendees of your presentation and the Q&A session. I typically inquire about this with the hope that participants will have an opportunity to provide input anonymously and I will receive the results.
- Supply of your business cards.
- Placement of podium to the left of the screen, as viewed by the audience, if PowerPoint or similar will be used.

Consider the last item. Most of the world's languages are written and read from left to right—some, such as Arabic, Hebrew, and Japanese, go from right to left. Therefore, when you, as an English speaker, mention something and then show it on the screen in text or as an image, most audiences will probably be most comfortable looking at and listening to you and then shifting their view to the right—to the screen. They prefer to "read" you and your slides from left to right [12, 13].

Actors and others who work in theater call the indicated podium position, relative to the screen location, stage right. That is, from the perspective of an actor on the stage, the podium is stage right of the screen.

A final logistics thought. Regardless of whose computer you plan to use, back up your files on a thumb drive that you keep on your person, in case you need it.

4.5 SPEAKING ADVICE: PRESENTING

4.5.1 Foil Fear

As we turn to the presentation part of the three-part speaking process, some of us will recall Section 4.1.1's mention of glossophobia, the fear of public speaking. If that includes you because you will speak soon somewhere, consider some thoughts about glossophobia and some suggestions for dealing with it.

Realistically, most of us will experience some anxiety prior to speaking, which should not debilitate or markedly hinder us. Comedian Johnny Carson said he was nervous prior to every one of his nighttime television monologues—that's over 4000 times [14].

Perhaps you applied some of the advice in Section 1.5.5, such as joining a public-speaking improvement organization like Toastmasters International. On reflection, that one commitment has reduced your apprehension indicating that the direct approach, including doing what you fear, has already worked. More proactive efforts will yield additional positive results.

Maybe you at least skimmed the 11 speech preparation ideas in Section 4.4 and selected and applied some of them. Jack Valenti stated, "The most effective antidote to stage fright and other calamities of speech making is total, slavish, monkish preparation." Section 4.4 offers ways to move toward Valenti's level of preparation.

Recognize that, when you give that scheduled speech, you will be more likely to see audience members who disapprove of or oppose your message than those who are neutral or supportive. Called the anger superiority effect, this phenomenon reflects the evolved human need to detect threats [15]. If you give disproportionate attention to opposers over supporters, one or a few of the former could lead you to conclude that you failed to achieve your speaking purpose, even though the vast majority of audience members were neutral or supportive.

Building a fear of an impending presentation requires that you imagine many things that could go wrong and, concluding that, if some do, you will produce a mediocre or failed result. Worse yet, you picture a perfect storm in which almost everything comes up short—a disaster.

In stark and positive contrast, you can instead imagine so many things that could go right, such as:

- you appeared and sounded cordial, confident, and competent
- most of the audience listened attentively, and many members revealed, through comments and body language, their understanding of and support for your message
- the Q&A session was spirited and informative
- you learned more about your topic as a result of interacting with the audience
- you left the event with firm plans to converse and maybe work with several session participants

Given that you have a choice, choose the second positive speech scenario, or something like it, and ignore the first negative scenario. As you begin to speak, think, and feel "as if" you are living that positive speech scenario, and you are likely to do so. Apply the advice of sports psychologist Bob Rotella who says, "The secret to great performances . . . is in the mind" [16].

Even with careful preparation, we will have speaking setbacks. I vividly recall sitting on a low stage, along with three other speakers and a moderator, just before the moderator was to start the session at an engineering conference in Milwaukee, WI. I was near the edge of the stage, and, you guessed it, my chair and I fell off the stage.

While that was decades ago, I recall the two major audience reactions—hurtful laughs and caring comments along the line of "Stu, are you OK?" I did what most of you would do, got up, put my chair on the stage, climbed up on the stage, and sat down. Later, after an introduction, I gave my presentation, which was not my best. The point of the story is that there was no way I was going to allow this careless and embarrassing event to kill my commitment, as noted in Section 1.2.2, to continuously improve my communication KSA.

Finally, put yourself in your audience. You and others want you, the speaker, to succeed. Most listeners have been in your shoes—they understand your situation. Furthermore, they want to make effective use of their time by listening to and learning from your message.

4.5.2 Verify Logistics

Arrive early and verify, including connecting and testing equipment, the relevant items listed in Section 4.4.11. Don't assume that various devices will work simply because they are there. Recognize that personnel who set up presentation rooms, such as at a convention center or hotel, typically do not give presentations. Therefore, you may need to make some adjustments—I always do. One or more of those personnel may still be there and able to assist you with your final preparations.

Find and test light switches. Sit in all four corners of the seating area and determine if you would be able to see the speaker and slides or props. Locate the restrooms; you may be asked.

4.5.3 Connect with Audience Members

Before You Speak

As you verify logistics, audience members will begin to arrive, as well as the host or moderator. Think about your knowledge of the audience you are about to speak to, and use that information to connect with individuals. Mingle a bit, introduce yourself, inquire about interests and concerns, encourage participation in the Q&A session, and maybe exchange business cards. This initial connection will enhance communication between you and them when you speak.

While You Speak

Continue to connect preferably immediately after being introduced. Share a startling statistic, ask a thought-provoking question, read a short poem, conduct a quick show-of-hands survey, read a quote, tell a brief story about yourself, or share a word that is new to you [3].

For example, when I began to speak about how to be more innovative at the annual conference of the National Society of Professional Engineers (NSPE), I mentioned that I had learned a new word. It was "metacognition," which means thinking about how we think [17]. If I were to speak to elementary school teachers, I would immediately connect with them by fondly mentioning Ms. Beth Blaha, the grade school teacher who introduced me to geology.

As you speak, consider doing some of the following to stay connected throughout your presentation:

1. Exhibit enthusiasm as you share useful facts and sincere emotions. Offer what's on your mind and in your heart, especially when referring to topics clearly related to the purpose of your presentation.
2. Use notes, which is clearly acceptable, if you frequently connect with the audience.
3. Vary volume, raise your voice, and at other times, speak quietly, slowly, or pause to encourage the audience to contemplate what you just said.
4. Speak loud enough—don't make audience members work to hear you. Some speech coaches think that the use of smartphones and other electronic devices, where volume is not an issue because it is adjustable, causes us to speak more softly [18].

5. Strike meaningless words, such as "ah," "um," "you know," and "currently," consistent with your practice session.
6. Repeat some words or phrases to emphasize them. For example, say, "With Option A we can put the satellite in orbit in half the time—half the time of Option B."
7. Occasionally move about—step to the side of the podium or maybe walk toward and even enter the audience. If you are showing slides, a remote control enables this approach.
8. Pass around some of the props that you use. Handling a prop enhances understanding of why you used it and strengthens the memory of the point you were making.
9. Use helpful body language (Section 2.4.3), such as an open stance contrasted with arms tight to the body or across the chest.
10. Try the previously mentioned "eye fives," that is, every now and then, talk eye-to-eye with one audience member for about five seconds.
11. Stop and take a time out if someone in the audience appears incredulous, offended, or angry. Diplomatically ask them to share their concern and suggest that it be the first topic discussed during the Q&A session.
12. Speak to the audience, not the screen, which typically puts your back to the audience. You might think that this suggestion is unnecessary. However, I have seen too many speakers spend too much time looking at and talking to the screen, appearing to ignore the audience. That is why I urge you to place the computer, used for any part of your presentation, between you and the audience (Section 4.4.11). Then you can look at and speak to the audience, occasionally glancing at your slides on the monitor while the audience sees the slides on the screen.

The preceding are not prescriptive actions that must be used at all speaking events. Instead, they are suggestions based on my experience and research, from which you may choose some to meet your needs. Consider the preceding and other lists as smorgasbords you scan and from which you choose whatever best suits your tastes.

4.5.4 Avoid the Need to Apologize

Nothing detracts more from a presentation than when a speaker begins by apologizing for something and then is later sorry for other aspects of the presentation. The speaker may apologize for lack of adequate preparation time, weak content, illogical organization of content, difficult-to-read slides, and exceeding allowable speaking and Q&A time. These and most other elements of a presentation are or should be under the speaker's control.

Perhaps some speakers apologize because they want to appear modest, in contrast to coming across as domineering. However, excess modesty can easily look like carelessness and incompetence.

Yes, sometimes we have to deal with circumstances that are out of our control. As noted in Section 4.4.7, I once lost the use of my computer while using PowerPoint during a presentation, and, another time, the internet connection failed during a webinar that I was presenting. Promptly apologize in such situations and do what you can to carry on or rectify the problem.

4.5.5 Conclude Definitively

As you near the end of your presentation and before inviting questions, briefly revisit your purpose or intended outcomes, indicate how you addressed them, and thank the audience for their attention. Possibly step out from behind the podium and maybe toward the audience as you say the preceding to indicate that you are finished speaking and ready for questions, suggestions, critiques, and other comments.

Some people say we tend to remember what we hear last, whether we are referring to the end of a sentence or the end of a speech. This is the stress position idea, where position means at the end [19]. Others argue for the primacy effect, claiming that we are most likely to recall what's at the beginning of a sentence or presentation [20].

Call it what you want, but recognize that the audience is most likely to remember the last words in a presentation. Don't dilute your principal message by also including secondary topics such as thanking various helpful colleagues, recommending follow-up studies, and referring to other related investigations. Address these topics in other ways, such as on one or more slides within your presentation, as part of a handout, or in the acknowledgments section or appendices of a written paper that delves into the details of your presentation.

4.5.6 Prompt Post-Presentation Question and Answer Session

You just completed speaking to students and faculty, other engineering practitioners, or some other audience. You concluded definitively, which reminded the audience of your purpose and how you achieved it. Your carefully prepared and practiced presentation helped you learn much more about your topic (as explained in Section 1.3.3). Many audience members who watched and listened attentively also learned. Some were neutral, others seemed to welcome and support your message, and a few apparently disapproved of or opposed it.

Some seem primed to share views or ask questions. You, as a student or practitioner, can orchestrate more thinking and learning by encouraging questions, suggestions, and other comments and doing your best to answer or otherwise respond to them.

4.5.7 Getting the Question and Answer Session Started

Sometimes we encounter a reluctance to ask the first question. We need that critical question in order to start a Q&A domino effect, a chain reaction. Assume you think this problem might occur. Then, prior to the presentation, ask a friend or colleague in the audience to ask a question, if needed, to start a chain reaction. Do not prescribe the question.

If no one offers a question, then be prepared to say honestly something like this: "The last time I spoke about this topic, someone asked. . .and my response was. . ."

Another approach is to look for someone in the audience who appears inquisitive or perplexed. Walk gently toward that individual and say something like, "you look concerned." My experience is that they will offer a thought—because most of us dislike the vacuum created by silence and, therefore, want to fill it. You just tipped the first domino.

4.5.8 Responding to Different Kinds of Questions

Once the Q&A is underway, restate questions if, for whatever reason, you suspect that the audience cannot hear them. Another reason for restating some questions is to move from a negative tone to a more positive one. For example, you, a consulting engineer, just finished speaking about your services, and someone asks, "Why are your services priced so high?" To be more positive, you might restate the question as "How did we arrive at the cost of our services?" [13]. You may also need to restate a question, or ask a question about it, to make sure you understand it before trying to answer it.

If you do not know the answer to a question, ask if someone in the audience does. Your audience may include many experts. As an engineering dean, I often led the hosting of groups of prospective students and their parents. I always invited some of our current students to join us—and referred many questions, especially those I could not answer as well, to them. Our students were experts on many topics of interest to our visitors.

Another approach to responding to questions, for which you do not have the answer, is simply to say so and try to be of some help. Ask the asker for a business card or other contact information and promise to follow up.

Not knowing all the answers often sets you up for learning. I once spoke about ways to help engineers be more creative and innovative and advocated experimenting with the use of freehand drawing—I have benefited from it. One audience member asked if listening to music could help. My response: I don't know. I walked away from that Q&A session with the names of two books about connections between music and creativity and innovation.

Consider a thought about body language when responding to questions, especially difficult ones. While listening to the question, stay put or move toward the asker. Do not back up, that is, step backward, which the audience may interpret as weakness or lack of preparation. Instead, focus on, answer, and engage the questioner.

If you are strongly or aggressively challenged, respond in an unemotional manner. That is, ignore the negative tone, acknowledge the comment, and respond by focusing on facts.

Finally, as the Q&A winds down and the speaking engagement ends, I suggest not being in a hurry to leave, unless you have to because of a scheduled subsequent use of the room [3]. Individuals, who are reluctant to comment in a group setting about a sensitive or personal issue, may approach you now—such post-session discussions could be mutually beneficial.

Once, as an audience member, I approached the speaker after the session concluded, complimented him on his presentation, and noted that we appeared to have some common colleagues. One thing led to another, and I soon entered into a multiyear mutually beneficial consulting arrangement with the speaker's firm.

4.6 SPEAKING ADVICE: FOLLOWING UP

Assume that your presentation, including the Q&A, went well. However, as suggested in Section 4.3, your speaking will be most successful—accomplish its purpose—if you view it as a three-part process. Therefore, we turn to the third part, following up.

4.6.1 Say Thank You

If you followed much of the advice offered in this chapter, you made a major effort to prepare and present. However, you probably did not act alone. Whether you are a student or practitioner, now may be the time to think about and thank those who helped you, such as [3]:

- Your professor, supervisor, or friend who knew you had something to talk about, encouraged you, and perhaps helped you "get on the program."
- Students, colleagues, and others who observed your practice sessions and offered insightful comments, provided content, attended your session and made more suggestions, or helped in other ways.
- The person who hosted or moderated your presentation and on-site personnel who assisted in various ways.

A brief text, email, conversation, or phone call could be meaningful. Consider a handwritten thank-you note; handwriting personalizes your message. Because it takes extra effort and is rare, the recipient is more likely to appreciate and remember your thoughts.

4.6.2 Commit to Improving

Recall if session participants anonymously evaluated your presentation. Maybe someone sat in on your presentation—as my wife often has—to assess audience reactions. Perhaps you or someone else made an audio or audio-visual recording. Obtain and review all of the inputs.

Experience suggests, as illustrated in Figure 4.3, that simple little changes in what you say and do result in large improvements in your speaking effectiveness.

What do I mean by simple changes? Consider these examples:

- Eliminate verbal graffiti such as "ah," "um," "currently," "you know," and "to be honest" (so you've been lying up to now?!). You can practice this improvement in your many daily conservations.
- Stop fiddling with eyeglasses or keys.
- Look at the audience, not the screen.
- Use a larger font size and fewer words on slides.
- Increase vocal variety.
- Move around—don't stay glued to the podium.
- Don't cross your arms on your chest during Q&A.
- Finally, ask yourself if you focus enough on the audience. The screen, your notes, the podium, the ceiling, or the floor don't make decisions, retain consultants, approve budgets, or support suggested changes [12].

You can readily make most of these simple changes on your own—provided you know they need attention.

Some moderate—take more effort—changes would be practicing your presentation out loud multiple times, completely redesigning your slides, and using

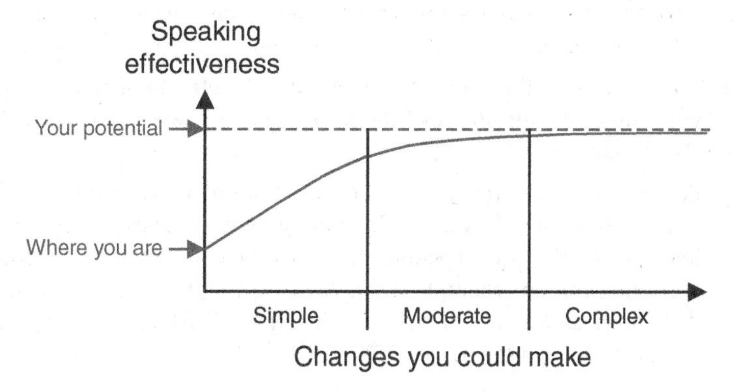

Figure 4.3 Little changes produce big improvements in speaking effectiveness.

props. A major change would be hiring and working with a speech coach. Figure 4.3 reminds us that simple little fixes, provided we recognize the need for them, produce large speaking improvements.

4.6.3 Keep Promises

Recall the purpose of your presentation. Think about those few audience members you met, briefly talked with, and exchanged contact information. Follow up on your promises, such as answering a question, sending an article, and providing information about a book. The ultimate success of your presentation may be a connection to and ongoing interaction with as little as one person who is in that audience and is capable of helping you achieve your purpose and advance your cause [3].

4.6.4 Leverage Your Presentation

You have invested a major effort in preparing, presenting, and following up on your presentation. Imagine that you are a talented sculptor who just created the greatest work. Would you and the world be properly served if you stored the statue in a dark basement room?

You just created what may be your greatest speech. Don't hide it. Instead, share it and build on it. Enable a successful presentation to help you, your university or employer, engineers, and others.

Some leveraging ideas [3, 21]:

- Make the presentation for other audiences, tailored to their interests. If you, as an engineering student, originally spoke to students and faculty in your university department or college, students and faculty in other campus departments or colleges may be interested. If you, as an engineering practitioner, spoke to personnel in your office, those who work in other of the firm's offices may want to hear your message, which you could deliver F2F or virtually.
- Speak again to the same audience, but now about a different topic that you think would interest them because of what you just learned about them.
- Prepare an article, based on your presentation, and submit it to a journal or magazine for publication. Content that you successfully presented to a live audience of say 50 people could now be enjoyed by 5000, 50,000, or more readers. For example, *Civil Engineering*, the flagship magazine of the American Society of Civil Engineers (ASCE), reaches 150,000 civil engineers around the globe [22].
- Create a presentation or an article that includes new developments or focuses on other aspects of the original presentation. An example of the

latter: You originally described a new approach to design, and now you will discuss manufacturing or construction implications.

- Mention your presentation on social media and offer to send the PowerPoint PDF file or a written version to any student, practitioner, or other person who requests it. Indicate the possibility of a virtual or F2F version tailored to a new audience.
- Insert a copy or summary of your presentation into your company's statement of qualifications, proposals, and other marketing materials available on your company's website.

4.7 EXAMPLES OF EXCELLENT SPEAKING

In this chapter, I offer ideas and advice to help you more effectively prepare, deliver, and follow up on speaking opportunities. As I completed an early draft of the chapter, I asked myself how else I could assist a reader who aspires to be a good to excellent speaker. My answer, after some pondering, was to introduce interested engineering students and practitioners to excellent speakers and their speeches. Accordingly, I offer the following sample of great speakers and speeches drawn from a variety of professions, occupations, and situations:

- President Abraham Lincoln's "Gettysburg address" delivered on November 19, 1863, at what was then the Soldiers' National Cemetery in Gettysburg, PA
- Engineer Herbert Hoover's "rugged individualism" speech presented on October 22, 1928, in New York City, near the end of his successful presidential campaign
- British Prime Minister Winston Churchill's "we shall fight on the beaches" speech delivered to the House of Commons in London on June 4, 1940
- President John F. Kennedy's "go to the moon in this decade" speech presented at Rice University on September 12, 1962
- Reverend Martin Luther King's "I have a dream" speech at the Lincoln Memorial in Washington, DC, delivered on August 28, 1963, as part of the March on Washington
- Nobel Peace Prize Winner Malala Yousafzai's "every child should receive an education" speech delivered in Oslo, Norway, on December 10, 2014, on acceptance of the prize

Consider googling some and seeing what more you can learn about speaking—especially content and delivery.

4.8 KEY POINTS

- Speaking, the most challenging of the five modes of communication, means using spoken language to convey information or feelings to audiences in business, government, academic, and similar settings
- Effective speaking can benefit the speaker, audience, speaker's employer, the engineering profession, along with the public it serves
- Speaking, like writing, enhances life-long learning
- While writing is the best way to transmit data and information, speaking is the best way to advocate markedly new approaches
- This chapter offers about two dozen immediately applicable forms of speaking advice organized into three categories—preparing, presenting, and following up
- Studying great speakers and their speeches helps us learn more about content and delivery

---

The notes I handle no better than many pianists.
But the pauses between the notes—
ah, that's where the art resides.

—*Artur Schnabel, musician and teacher*

EXERCISES

4.1 Speaking improvements you would make: Select a presentation you recently made, and don't necessarily limit it to engineering. Based on the speaking advice shared in this chapter, describe up to three things you would change.

4.2 Evaluating speaking advice: Review Sections 4.4, 4.5, and 4.6 and determine what the most valuable advice is for you right now. Indicate what you selected and why.

4.3 Out-loud practice: If you already regularly use comprehensive out-loud speaking practice, as described in Section 4.4.10, perhaps the instructor will excuse you from doing this exercise. For others, identify a near-future speaking opportunity. As your content nears completion and you have a first draft, conduct at least two or three out-loud practices of the entire presentation and additional practices with the most challenging parts. Make improvements to the content after each practice. Consider making an audio or audio–visual recording of your trial runs, or having someone observe and comment, so that you can fully hear, see, and learn from your presentation. Give the speech and then share, in writing, your views about the pros and cons of out-loud practice.

4.4 Learning from a widely known speech: Find what you think was an influential—preferably in a positive manner—speech. To help you get started, consider the speakers and speeches listed in Section 4.7. If feasible, obtain and work from an audio-visual version as opposed to a transcript. Study the speech and then describe its most positive and negative features.

REFERENCES

1. Black, R. (2019). Glossophobia (fear of public speaking): Are you glossophobic? Psycom. 12 September 2019. https://www.psycom.net/glossophobia-fear-of-public-speaking (accessed 1 June 2023).
2. Encyclopedia.com. (2023). https://www.encyclopedia.com/humanities/dictionaries-thesauruses-pictures-and-press-releases/speaking (accessed 17 May 2023).
3. Walesh, S.G. (2012). *Engineering Your Future: The Professional Practice of Engineering*. Chapter 3, Communicating to make things happen. (pp. 97–98). Hoboken, NJ: Wiley.
4. Walesh, S.G. (2016). Using neuroscience to work smarter: more effective, creative, innovative. Third U.S.–Japan Geoenvironmental Engineering Workshop. ASCE. Chicago, IL. August 13–14.
5. Valenti, J. (2002). *Speak Up With Confidence: How to Prepare, Learn, and Deliver Effective Speeches*. Chapter 4, The delivery. New York: Hyperion.
6. Crunchbase. (2023). PitchVantage. https://www.crunchbase.com/organization/pitchvantage (accessed 14 September 2023).
7. PitchVantage. (2023). Universities. https://pitchvantage.com/how-it-works-universities/ (accessed 14 September 2023).
8. Decker, B. (1992). *You've Got to be Believed to be Heard*. New York: St. Martin's Press.
9. Walesh, S.G. (2004). *Managing and Leading: 52 Lessons Learned for Engineers*. Lesson 23, Practice out loud. Reston, VA: ASCE Press.
10. Selbert, P. (2006). Missouri museum recalls historic Churchill address. *Chicago Tribune* May 21.
11. Reynolds, G. (2008). *Presentations: Simple Ideas on Presentation and Delivery*. Berkley, CA: New Riders.
12. Koyfman, S. (2012). Why is most language read from left to right? *Babbel Magazine*. April 28. https://www.babbel.com/en/magazine/right-to-left-languages (accessed 29 May 2023).
13. Cooper, C. (2004). A winning presentation. *Investors Business Daily*. July 1.
14. Koegel, T.J. (2007). *The Exceptional Presenter*. Austin, TX: Greenleaf Book Group Press.
15. Restak, R. and Kim, S. (2010). *The Playful Brain: The Surprising Science of How Puzzles Improve Your Mind*. (pp. 235–238). New York: Riverhead Books.
16. Burke, M. (2007). How not to choke. *Fortune*. November 26.
17. Walesh, S.G. (2018). Avoid being stung by einstellung: then innovate. Presented at the annual conference of the NSPE. Las Vegas, NV. July 18–22.

18. Parkinson, J.R. (2018). The other side of listening. *Herald-Tribune*. Sarasota, FL. March 3.
19. Anholt, R.R.H. (2006). *Dazzle 'Em with Style: The Art of Oral Scientific Presentation-Second Edition*. Amsterdam: Elsevier.
20. Irish, R. and Weiss, P.E. (2009). *Engineering Communication: From Principles to Practice*. Ontario, CA: Oxford University Press.
21. Walesh, S.G. (2004). *Managing and Leading: 52 Lessons Learned for Engineers*. Lesson 22, Preparing, presenting, and publishing professional papers Reston, VA: ASCE Press.
22. Civil Engineering Source and Civil Engineering. (2023). https://www.asce.org/publications-and-news/civil-engineering-source/civil-engineering-magazine/issues/about-civil-engineering-magazine (accessed 31 May 2023).

CHAPTER 5

USING VISUALS

> Vision trumps all other senses. . .
> taking up half of our brain's resources.
>
> —*John Medina, biologist and author*

After studying this chapter, you will be able to:

- Explain why engineers should frequently use visuals in their written and spoken communications
- Provide examples of two kinds of visuals; images and props
- Discuss engineering's advantage, relative to most occupations, in the use of visuals
- Use some of the visuals' advice to prepare more effective documents and speeches or to help others do so
- Identify benefits to engineers of doing visual arts

5.1 VISUALS DEFINED

The Merriam-Webster dictionary defines a visual as "something (such as a graphic) that appeals to the sight and is used for effect or illustration—usually used in plural" [1]. In a similar manner, the Cambridge dictionary says a visual is "something such as a picture, photograph, or piece of film used to give a particular effect or to explain something" [2].

For this book's purposes, and based largely on those two similar definitions, a visual is something appealing to the sense of sight that we use when speaking

The Communicative Engineer: How to Ask, Listen, Write, Speak, and Use Visuals, First Edition.
Stuart G. Walesh. © 2024 John Wiley & Sons, Inc. Published 2024 by John Wiley & Sons, Inc.

or writing to inform, convince, or otherwise engage others. That may sound like a tall order. Fortunately, as explained in the next section, vision is up to it because it is our most powerful sense.

5.2 VISION: THE MOST POWERFUL SENSE

We use our five basic senses of vision, hearing, smell, taste, and touch to gather and act on information about the world around us. Vision is our most powerful sense because it uses more of our brain than any other sense. Therefore, whenever you communicate, whether speaking or writing to one person or a large audience, consider using visuals to engage more fully the brains and hearts of the other person or audience members. They will be more likely to understand, remember, and act on your messages [3].

As the adage, whose essence is popularly attributed to the Chinese philosopher Confucius, says, "A picture is worth a thousand words" [4]. I devote a chapter to vision because it is the most powerful sense and rarely receives explicit, detailed, and pragmatic attention in communication books, especially contrasted with writing and speaking.

One way of illustrating the importance of vision is to note how often visual words appear in our writing and speaking. We *see* someone's point of *view*, *envision* a *bright* future, fear *dark* alleys, *reveal* our *blind spots*, promise to be *transparent*, and share an *insight* [5].

Another way of showing the importance of vision is to learn how one wise engineer, Taiichi Ohno, employed vision to improve the processes used to manufacture Toyota vehicles. He arranged for new engineers and managers to go to a visually advantageous location in a factory and draw a two-foot-diameter circle on the floor. They were asked to stand in the circle for up to eight hours, observe, and then share what they learned and the improvement ideas that occurred to them. Engineer Ohno's intent in having new personnel participate in the Ohno Circle was to get them beyond sitting in an office and looking at data about the manufacturing process. Instead, they went to the process and made deep observations [6, 7].

Yes, eight hours would be a long time. However, what if you or I devoted 30 minutes at an unobtrusive location and observed a variety of processes and places, such as construction sites, manufacturing lines, meetings, laboratories, reception desks, cafeterias, classes, and loading docks? I am confident that we, without special relevant expertise, would see some underutilized resources, excess motion, defects, excess inventory, non-value-added steps, waiting, and/or safety and health hazards [7]. Like baseball great Yogi Berra said, "You can observe a lot by watching" [8].

5.3 EXAMPLES OF VISUALS

Simply put, a visual is something that we show to an individual or an audience to engage them cognitively and emotionally. Within the engineering world, we can draw on a huge and varied supply of visuals. Some examples are freehand sketches, formal drawings, graphs, icons, photographs, flow charts, tables, symbols, slides, display boards, videos, and props.

"Prop," the last visual listed," comes from the theatrical world and is a shortened form of "property," which means any object handled by an actor during a performance. When you and I give a presentation, we are like actors giving a performance during which we strive to communicate with our audiences. Props may help us.

For example, I spoke to engineering students about "10 Tips for Success and Significance." For each tip, I reached into a box and pulled out an understandable and memorable prop, with some examples being a dollar bill, gold pocket watch, books, mortarboard, and crystal vase. Figure 5.1 shows all of the props.

Figure 5.1 Props helped to explain each tip in the presentation "10 Tips for Success and Significance."

The crystal vase represented a person's reputation—protecting it was one of the tips. Each person's reputation, like a handcrafted vase, is unique. Major time and effort go into building a reputation and creating a crystal vase. Once shattered, a reputation, like a vase, is impossible to restore.

Recall the auditory, visual, and kinesthetic learners described in Section 1.5.3. Props are especially effective when communicating with visual and kinesthetic learners because they can handle props and view them from various perspectives.

For discussion purposes, let's define all of the above-listed visuals, minus props, as being two-dimensional (2D) and then refer to props as three-dimensional (3D). I think props are more effective for communication than other visuals because their 3D status means that they are the real thing or a model of it. 2D visuals are usually just images of the real thing. Let's be creative and use whatever visuals work, including props.

5.4 ENGINEERING'S EDGE IN USING VISUALS

We engineers and other technical professionals plan, design, construct, and operate things—products, processes, structures, facilities, and systems—that serve societal needs. We produce results that are figuratively, and sometimes, literally concrete. Accordingly, we have ready access to highly varied potential visuals, including props, the use of which enhances communication with those we serve. If you are going to talk about something, to a person or group, you usually have the options of using an image of the "something," the actual "something," or a portion of it.

5.5 SUGGESTIONS FOR USING VISUALS, EXCLUDING PROPS

Before reading this section about using 2D visuals and the next section about using props, I suggest re-reading Section 1.5, "Principles of Effective Communication." The first three of the six principles—know audience, state purpose, and accommodate preferred ways of understanding— apply to the use of most visuals.

5.5.1 Consider PowerPoint and Beyond

Commit to being open to using varied 2D visuals. Stated differently, don't immediately assume the application of PowerPoint or other slideware. These visual tools offer many attractive features, such as animation, but are used so frequently that they sometimes evoke visual fatigue—"not another PowerPoint!"—instead of drawing the audience to you and your topic. Consider other 2D options such as a newsprint pad, display boards, white boards as in a classroom, or handouts.

I was part of a team of engineers that officially released a report at the National Academy of Engineering (NAE) in Washington, D.C. We used, in series, five display boards and, after discussing each, left it and the others on display behind the seated speakers for later reference. From the audience's perspective, our presentation evolved one board at a time, and when we were finished speaking and ready for the question and answer (Q&A) session, the audience could readily see all parts of the presentation. Board use was effective judging from the productive Q&A session.

5.5.2 Use the Most Effective Image Format

Topic-subtopic Slides and Other 2D Visuals Are Least Effective

Minimize use of slides and other 2D visuals that use topic-subtopic slides, that is, a statement followed by bulleted points, as illustrated in Figure 5.2. You, not your audience, would be the principal beneficiary of this kind of slide because you could use it as your notes when you speak. However, because you would not want to read the slide to them, you would probably ad lib.

Now, while you are improvising, imagine yourself in the audience and what they are seeing, hearing, and doing. They read one set of words, hear you saying another set, and get confused. Despite your good intentions not to read to the audience, many audience members probably wish, to avoid inconsistency, that you simply read the text. That raises another issue—why are you needed?

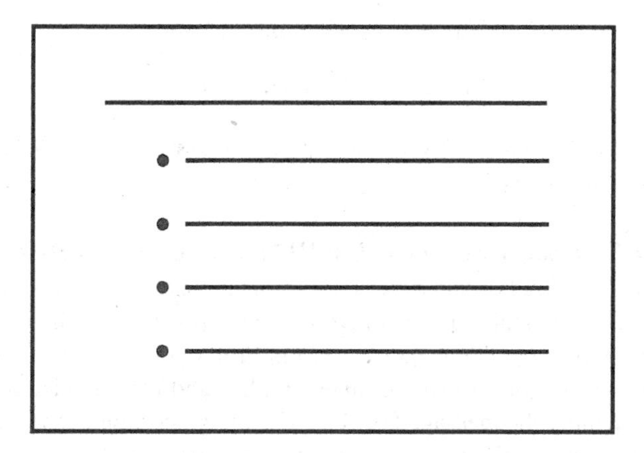

Figure 5.2 This frequently used format for slides or other 2D visuals may aid the speaker but is likely to confuse the audience.

Use of topic-subtopic slides is encouraged by default settings—easy-to-use templates—in PowerPoint and similar slideware. Various rules-of-thumb also drive the use of bulleted slides and 2D visuals. For example, the 1-6-6 Rule states that each visual should have one main idea, a maximum of six bullet points, and a maximum of six words per bullet point [9]. Easy to do and may make intuitive sense in that it will restrict the number of words on bulleted visuals. However, such rules, because they contain "rule," may encourage even more visuals that are bulleted.

Sometimes bulleted slides, that is, all text, strengthen parts of a presentation. Figure 5.3 is based on a slide I used near the beginning of the American Society of Civil Engineers' (ASCE) webinar "Speaking: How to Prepare and Deliver a Convincing Presentation." The slide succinctly states the intended webinar outcomes, which were revisited near the webinar's end.

Incidentally, ASCE sponsored the webinar. Therefore, the text on slides was mostly deep blue, with some red for effect, because that blue shade is the dominant color in ASCE's logo and on its website. My emphasis on society's major color provided another connection between the audience and me. More on the use of color later (Section 5.5.5).

I prepared the original Figure 5.3 before learning the 1-6-6 Rule and inadvertently produced a slide that conforms with the rule. Another way of saying that, if we want to create a topic-subtopic slide, we should minimize the number of bullets and words.

As a result of this webinar,
You should be able to:

Explain the power of speaking

Apply speaking tips

Make more things happen

Figure 5.3 Effective bulleted PowerPoint slide is used to state the intended outcomes of a speaking improvement webinar.

Assertion-Evidence Slides and Other 2D Visuals Are Most Effective

So what works? Studies and experience indicate that the most effective slide, or other 2D visual, contains a declarative statement (assertion) with or followed by a supporting image (evidence), as shown in Figure 5.4. Audience members read the statement, see the supportive image, understand the intended meaning, and tend to remember the message [10, 11]. The image portion of the slide, or other 2D display, engages the powerful vision sense and, therefore, enhances understanding and remembering. Assertion-evidence slides also benefit the preparer of them.

Figure 5.4 The most understandable and memorable 2D image consists of a short declarative statement followed by a supporting image.

Beneficiaries of Assertion-Evidence Slides

The previous paragraph briefly stated that assertion-evidence slides benefit both the audience and the speaker who prepares the visuals. Let's explore this. Researchers Garner and Alley [12] summarize studies in which some students received a lecture during which topic-subtopic slides were used and another group received the same content, but now supported by assertion-evidence slides. When subsequently tested, students who participated in the assertion-evidence lectures "had superior comprehension and superior recall."

Now consider the slide preparers. Preparing effective assertion-evidence slides or other 2D visuals requires the preparer to study a topic, or portion of it, for presentation in 2D; select key ideas and facts; summarize them in a short sentence; and choose a supportive image. This text and image thought process requires more effort—more learning—per slide than listing a topic and its subtopics [12].

You may say, I want to convey more ideas or information, which I can on a word-rich bulleted 2D visual. Resist the temptation to pile on more content. Instead, design a series of effective slides or 2D visuals, most of which use a declarative statement and supporting image, that progressively introduce your ideas and information. Stated differently, keep each display simple for ease of understanding and accomplish communication of complexity with multiple effective slides.

Image-only Slides

Having discussed 2D visuals that are all text and some that contain text and an image, the remaining option is a visual that is simply an image—no text. For example, you could be creating the visuals for a presentation, during which you will stop speaking at the midpoint and ask participants for questions or comments. Therefore, you insert a slide showing a question mark or a sport's referee making a time-out signal.

5.5.3 Favor San-Serif Fonts

I suggest using san serif fonts on 2D visuals, where "san-serif" means fonts without small strokes or fine lines projecting from the main strokes of a letter [13]. Some font examples follow:

Serif fonts: Bookman Old Style, Cambria, Times New Roman

San-serif fonts: Arial, Calibri, Tahoma

San-serif fonts have less clutter and, therefore, are easier to read on 2D visuals. In contrast, serif fonts seem to be easier to read in extensive text, such as in this book. The widely used Times New Roman font is an exception because, although it is a serif font, it is one of the easiest fonts to read in text and on visuals.

Font impact is subjective and you may prefer a serif font. If so, I suggest, in the interest of minimizing clutter, that you minimize mixing fonts on individual 2D visuals or even from one visual to the next.

5.5.4 Apply the 1/30 Rule

When selecting the size of alphanumeric characters for use with PowerPoint or other slideware, I find that a character height of at least 1/30 the height of the slide works well. This rule of thumb means using a 20pt. or larger font. Most members of an audience will be able to easily read the slide.

If you need some insight into how well an audience member will be able to read one of your slides that includes text, print it in the full-page slide format and place it on the floor. Then stand up and try to read it. If you can't comfortably do so, your audience will encounter the same obstacle when looking at that slide projected on a screen [14].

Sometimes, when preparing slides for a presentation, we find the perfect diagram in a magazine, journal, book, or elsewhere. We could easily use the diagram, with proper credit, by scanning it and creating a slide. Before doing that, confirm that the 1/30 height rule applies. Because the image was designed to be read close up, it is likely to violate the rule. One option is to create a handout of the diagram, showing the source, that you distribute prior to or during your presentation. Another approach is to create a slide that greatly simplifies, but captures the essence of, the original diagram.

5.5.5 Choose Meaningful Colors

Recall the earlier comment (Section 5.5.2) about preparing a slide for an ASCE webinar and using deep blue as the dominant color because that is the principal color in the ASCE logo and on its website. The idea is to use color as a way of connecting you to the organization that is sponsoring or arranging your presentation.

You can be confident that ASCE and other engineering societies carefully select the colors that represent them. Why? Because colors have meaning, including evoking emotions. Interestingly, most of the major engineering societies use blue, in many different shades. Blue's meanings include authority, stability, and wisdom.

The meaning of any color will vary among cultures and from person to person. Because colors have meaning, we should try to choose them to enhance our communication, especially in the writing and speaking modes. Use a color or colors consistent with the purpose of your presentation or document, and avoid counter-productive colors. Because of the subjective meanings of each color, let us not be overly optimistic about how widely our interpretation of color for text or an image will be shared.

Table 5.1 provides color interpretations based on sources and my experience [9, 15, 16]. Consider using it to start your selection of colors when you are preparing a 2D visual and when interpreting 2D visuals prepared by others. The communicative engineer benefits from color when sending messages and considers color when receiving messages.

In addition to using colors to evoke certain meanings, we can also use color to attract viewers and readers. For example, if you are drafting a report and it will eventually be available electronically, you could use color on the cover and within the document, such as having colored photographs, graphs, and line drawings.

A final thought about colors selected for use on 2D visuals. Informed by experience, I know that some viewers cannot readily distinguish certain color combinations. An example is red text on a blue background, or vice versa.

Table 5.1 Proactively use colors because they convey many meanings and, therefore, enhance our communication.

Color	Meanings
Black	Depression, grief, penitence
Blue	Authority, serenity, sincerity, stability, sadness, truth, wisdom
Green	Environment, growth, health, healing, greed, money, youth
Orange	Amusement, attention-grabbing, energy, extroversion, strength
Pink	Femininity, immaturity, love, softness
Purple	Dignity, royalty, spirituality
Red	Courage, danger, love, lust, rescue
White	Cleanliness, faith, innocence
Yellow	Cowardice, future, joy, nostalgia, treason

Prudence suggests testing your slides, preferably with the help of one or more other people, by showing them on a screen or viewing them on your computer.

5.5.6 Create Visual Analogies

Draw on the dominance of our sense of sight and create analogies to help introduce or explain concepts, ideas, information, and quantities to various audiences. Consider an example.

Assume you will be speaking at a public meeting about a proposed stormwater retention pond and want to describe the maximum volume of water in the pond. You might decide to use acre-foot to define the size of the 150-acre-foot pond, where one acre-foot is equivalent to a foot of water on an acre. However, you realize that many citizens cannot easily visualize an acre.

Therefore, you show an aerial photograph of a football stadium, note that the playing area is approximately one acre, and state that an acre-foot of water would be one foot of water over that playing area. Now most audience members can visualize 150-acre-feet of water. To reiterate, the analogy is that one acre-foot of volume is equivalent to one foot of water on a football field.

Audiences tend to understand aerial photographs of football fields. More importantly, they are likely to remember the message communicated by that visual analogy.

5.5.7 Transition From One Slide to the Next

Assume you have a draft set of slides or other 2D visuals. They will help you tell a story that supports the purpose of your presentation. Begin to think how you will bring the audience along with you—maintain their interest—as you transition from one slide, board, or other 2D visual to the next. Avoid showing a series of visuals with the following kind of repetitive and boring narrative: "This slide shows. . . , This slide shows. . . , etc." Instead, use various kinds of informative transition statements.

Consider an example. I was speaking to engineers about how we can be even more effective communicators. Early in the presentation, I shared my view that communication is essential to successful engineering. I showed a slide of an engineering dean who, years ago, urged me and other engineering students to learn how to write, speak, use visuals, and mathematics—He said communication, including how to apply those modes, was essential in engineering. That event was my introduction to the importance of communication in engineering.

Then, as part of transitioning to the next slide, I said something like this: "Another way to illustrate the importance of communication is to consider the consequences of poor communication." The next slide showed a collapsed structure, which I used to note the number of people injured and killed and explain how miscommunication caused the disaster.

Continuing, my communication-is-essential theme, I transitioned to the next slide by saying something like, "Consider a closer-to-home example of the importance of communication." Using a slide, I illustrated how an engineering firm lost clients mainly because of poor communication, not because of technical errors.

Bottom line: Select the point you want to make (e.g., communication is essential to successful engineering) and construct your argument (e.g., dean said so, prevent disasters, serve and retain clients). Then tell your story, verbally and visually, in an interesting manner, including maintaining audience interest as you transition to the next visual.

5.5.8 Provide Your Contact Information

Recall the handout discussion in the previous chapter (Section 4.4.7) which strongly suggests that the handout includes information about the speaker that enables audience members, or others, to easily contact the speaker after the presentation. More specifically, the handout should include information such as the presentation title, date, event, and sponsor, plus the speaker's name, credentials, title, employer, email, and maybe telephone number and website.

An effective way to do this, if you are using slideware in your presentation, is to provide the preceding information on the first slide, that is, title slide. Figure 5.5 illustrates the approach using a format similar to one ASCE and I developed to present many webinars.

This use of the title slide assumes you are going to provide a copy of the slides—electronically or in hard copy—typically as a portable document format

SPEAKING:

How To Prepare and Deliver
A Convincing Presentation

Presented 10 March 2021 by:

Stuart G. Walesh, PhD, PE, Dist.M.ASCE, F.NSPE

Consultant – Teacher – Author

stu-walesh@comcast.net

www.HelpingYouEngineerYourFuture.com

www.linkedin.com/in/stuwalesh

ASCE KNOWLEDGE & LEARNING

Copyright 2021 S. G. Walesh

Figure 5.5 Use the title slide to provide your contact information for easy use, after the presentation, by individuals interested in you, your topic, or your organization.

(PDF) file, before, during, or after your presentation. Including the sponsor's name recognizes their contribution, and the joint effort between you and them enhances the credibility of the presentation and the handout.

5.6 SUGGESTIONS FOR USING PROPS

I introduced and illustrated props, the 3D visuals, in Section 5.3. Props are effective communication aids because they maximize use of our sense of vision, the most powerful sense. Depending on what they are and how they are used—e.g., passed around the audience—props can also engage our other four basic senses of hearing, smell, taste, and touch.

Another way of recognizing the potential communication value of props is to recall the Section 1.5.3 discussion of preferred ways of understanding messages—auditory, visual, and kinesthetic. Props offer an effective way of communicating with all three types of learners, especially visual and aesthetic.

The best way to offer suggestions for using props is to describe actual uses by highlighting certain features and describing the communication result. Many examples follow.

5.6.1 Explaining Differences in Types of Pipes

As a consultant to a water supply utility, I chaired an advisory pipe panel charged with weighing the "pros" and "cons" of ductile iron pipe (DIP) versus polyvinyl chloride (PVC) pipe in parts of a water distribution system. The panel would report its findings to the utility's board.

The utility's construction manager opposed allowing PVC pipe. He made a presentation to panel members during which he offered his view of the pros and cons of DIP and PVC pipe.

He referred to a 20-foot-long, 8-inch-diameter PVC pipe that he had placed in the meeting room. He noted that when PVC pipe failed, the failure tended to affect the entire length of pipe and required an excavation more than 20 feet long. In contrast, DIP failures tended to be localized and, therefore, repaired by excavating a small hole down to and around the pipe and clamping a saddle over the failure. The differences in the lengths of pipe involved in repair meant that a typical PVC repair was much more costly than a typical DIP repair. The manager's use of props informed panel members.

After three months of study, the advisory pipe panel decided to recommend that the utility continue to use DIP, that is, not allow PVC pipe. Principal reason: DIP was the best way to serve the long-term interests of the city's infrastructure.

As part of the planning for the panel's meeting with the utility's board, we arranged to have a section of each type of pipe at the meeting. I wanted to be sure that all board members were familiar with the very different appearance, feel, and weight of these two materials. Board members connected with the props. My impression was that some had never seen or touched the DIP that made up the system for which they were responsible or the PVC pipe being proposed for use in that system. They were now better prepared to make decisions—they decided not to approve use of the PVC pipe because of what they saw and learned [17].

5.6.2 Illustrating the Cause of a Disaster

On Tuesday, January 28, 1986, the Space Shuttle *Challenger* lifted off the launchpad at the Kennedy Space Center in Florida, accelerating upward driven by the three main orbiter engines and the two rocket boosters. Seventy-three seconds into the flight, at almost 50,000 feet above the earth and moving at a speed of just under Mach 2, space shuttle *Challenger* exploded. The six male astronauts and high school teacher Christa McAuliffe, the first teacher in space, may have died immediately, or three or four minutes into the flight, when the crew compartment portion of the orbiter hit the Atlantic Ocean at 200 miles per hour.

The space shuttle included a pair of solid rocket boosters, also called solid rocket motors, mounted on either side of the fuel tank and small compared to it. The boosters were to provide two minutes of additional thrust beginning at lift-off, after which they would be jettisoned, fall into the ocean under control of parachutes, and be recovered for reuse.

Each booster consisted of four sections manufactured by Morton-Thiokol (MT) in Brigham City, Utah, and shipped by rail to Florida. During section assembly at the Kennedy Space Center, workers sealed the joints between the sections with pairs of rubber O-rings. The rubber O-rings are a key aspect of this disaster story.

Less than a day before the scheduled launch, a few MT engineers became concerned about the unusually cold temperature forecast for next morning's launch. More specifically, they concluded that the cold temperature would make the O-rings too rigid and gas blow-by could occur—that is, gas could flow from inside one or both booster rockets through joints and cause damage. They expressed their concern to MT management, but management, wanting to please its client, the National Aeronautics and Space Administration (NASA), decided to launch anyway. That decision led to the disaster.

The Presidential Commission on the Space Shuttle *Challenger* Accident (also called the Rogers Commission after its Chairman, William P. Rogers) worked for several months and published its report in June 1986. The commission concluded that O-ring failures, compromised by factors that included low temperature, caused the disaster. The report criticized both NASA and MT for not redesigning the joint and for poor communication and decision-making [18].

We turn now to a simple prop used during the work of the commission. Richard P. Feynman, a commission member, a physicist, and the winner of the 1965 Nobel Prize in Physics, focused on the previously mentioned O-rings. Noting that the air temperature at *Challenger's* launch was 28 to 29 degrees Fahrenheit, he conducted a simple experiment during a televised commission hearing and then reported the results. Using a clamp, he compressed a sample of the O-ring material, immersed it in ice water, took it out, removed the clamp, and noted that the material did not quickly rebound.

In his words: “I took this stuff that I got out of your seal and I put it in ice water, and I discovered that when you put some pressure on it for a while and then undo it, it does not stretch back. It stays the same dimension. In other words, for a few seconds at least and more seconds than that, there is no resilience in this particular material when it is at a temperature of 32 degrees.” Therefore, the O-rings were inadequate at similar low temperatures because they did not quickly rebound to their normal shape, “thus failing to maintain a tight seal when rocket pressure distorted the structure of the solid fuel booster” [19]. A classic example of prop use.

5.6.3 Describing the Human Brain and How to Use It Wisely

I have spoken to engineering students, faculty, and others about topics such as “Using Neuroscience to Work Smarter” and “You, Your Brain, and the Rest of Your Life.” Figure 5.6 shows props I used to illustrate some of my principal points. No PowerPoint is used or needed.

Figure 5.6 These props supported a description of the brain and how to use that knowledge to live and work smarter.

The cauliflower illustrates the approximate size and appearance of the human brain and then enabled me to describe its composition—such as 100,000 billion neurons—and discuss the different functions of the brain's left and right hemispheres. I used the gallon jug to explain that the human heart pumps blood carrying glucose, nutrients, and oxygen to the brain at a rate of up to 16 gallons per hour. The lightbulb provided a way for me to emphasize the small amount of power available to the brain—about 20 watts—and formed the basis for my advice to minimize multitasking because it misuses the limited power [3].

Having done many prop-based presentations, I offer the following observations:

- The box containing the props gets the audience's attention—most individuals seem to welcome something instead of another PowerPoint presentation.
- The audience stays connected because they wonder, "What will he pull out next?"
- Participants "get the message," as evidenced by the quick recall of the meaning of any prop near the end of my talk, as I refer to it again or put it back in the box.

5.6.4 More Examples of Props

I recall being at a meeting where an engineer explained various consequences of leaks in municipal water distribution systems. To illustrate one kind of damage, he brought a large, heavy brass valve to the meeting. It was deeply eroded—several inches—because of proximity to a water jet issuing from a hole in an underground water main. Seeing and feeling the damage caused by a water jet over a long time reminded me of much earlier standing on the rim of the Grand Canyon and seeing the work of water and wind over a long time.

A professor used a rectangular cross-sectioned foam beam, with longitudinal parallel lines drawn on it, to show tension and compression by bending the beam back and forth.

I advised students in my engineering management and leadership course to, when they begin their first job, invest ten percent of every dollar they make. They would retire as millionaires. I once illustrated this advice during a presentation by taking a dollar from my billfold, approaching a young engineer in the first row, and asking that person and others to imagine this was the first dollar earned as a graduate engineer. Then I extended the dollar to that student but pulled it back before the person grabbed it. I tore off about ten percent of the bill, handed the big piece to the engineer, and said, do what you want with this. Then I handed over the ten percent piece and said invest this.

As you prepare for your next presentation, think about ways you can prop it up with props. Talking about something may work, showing an image of it is usually better, and showing it, and maybe passing it around the audience, is best.

5.7 YOU ARE THE MOST IMPORTANT VISUAL

A word of caution. Having stressed the importance of visuals and offered many suggestions for using them, I must indicate that you are the most important visual—not images or props. You or I cannot "PowerPoint" or "prop" an audience into being aware of, understanding, promoting, opposing, financing, doing, or not doing, anything.

We must convey the message by using these three channels:

- Verbal—the words we use
- Vocal—how we use our voice
- Visual—our appearance and body language supplemented by the visuals we use

We had better be more excited about the purpose of our presentation than about the technology we use to present it. As someone said, if you can be replaced by PowerPoint, you should be!

Audience members will judge us, in part, by how we are dressed. Try to dress in a slightly more formal way than what you expect from the audience. They may collectively view a speaker who dresses "below" them as being disrespectful and poorly prepared. If, right before the presentation, you discover that you are overdressed, make adjustments such as taking off your sport coat or jacket, removing your tie or scarf, or rolling up your sleeves.

Refer again to Section 4.5.3 titled "Connect with Audience Members" for additional connecting ideas, some of which will determine how the audience "sees" you, before you speak and during your presentation.

5.8 DOING VISUAL ARTS

5.8.1 Change of Emphasis

Up to this point in the chapter, we have discussed how to use 2D and 3D visuals to improve our communication with others. I stressed leveraging vision, our most powerful sense, as we prepare visuals for use in documents, speeches, and other communication situations.

Now we shift from using visuals with the goal of communicating with others to creating visuals for pleasure and, for some of us, to increase our effectiveness as engineers. By creating visuals, I mean doing—not just reading about and viewing—visual arts such as drawing, painting, sculpture, and photography, which "meet the eye and evoke an emotion through an expression of skill and imagination." [20].

Note the focus on doing, not just studying, visual arts. My experience and research suggest that art's professional (and personal) value emanates from actively doing it. Incidentally, I suspect that some of what I share is applicable to the performing arts (e.g., see Root-Bernstein [21]).

Based on my knowledge of the engineering community, I know that some of this book's readers do visual arts—and some participate in performing arts such as music, dance, or drama. If you are an amateur or professional visual artist, my hope is that this discussion will broaden and deepen your understanding of the professional benefits of your art. Perhaps some other readers, after reading this in-depth discussion of visual arts, will explore doing visual art as an enjoyable diversion and maybe as an aid in their studies or professional work.

As for me, I am an amateur artist working in pencil and acrylic and preferring realism such as landscapes and animals, with Figure 5.7 being an example of the latter. After a five-decade lapse from freehand drawing instruction in grade school, and on a whim, I took a one-day freehand pencil-drawing course, made

Figure 5.7 Rescued Sara now has a home.

progress, and enjoyed it. Classes and drawing continued, initially as a diversion but also because, over a few years, I gradually saw intriguing connections between doing art and doing engineering (e.g., see [22]), some of which I am sharing with you in this chapter.

We, as engineering students or practitioners, can benefit professionally in at least two ways by practicing a visual art.

- First, we will experience enhanced vision—more seeing and less just looking
- Second, we will be even more effective composers, that is, spatial arrangers of ideas, principles, options, and objects

These professional benefits begin with the simple pleasure of completing a drawing, painting, sculpture, photograph, or other creation. Let's explore the two professional benefits—seeing more and composing effectively.

5.8.2 More Seeing, Less Just Looking

Recall the Section 2.4 discussion of how listening requires much more effort than hearing and how hearing, in turn, enhances communication because it enables us to obtain facts and understand feelings. In a similar manner, **seeing** requires much more effort than looking, and seeing, in turn, enables us to understand issues, problems, and opportunities more thoroughly and eventually resolve or pursue them.

Even though vision dominates, we often lack attention, a liability called "inattentional blindness." This "illusion of attention" means that "we experience far less of our visual world than we think we do" [23]. We may visually and otherwise miss much, especially creative opportunities. For example, why didn't Western Union invent the telephone, U.S. Steel the minimill, and IBM the personal computer? [24]. Doing visual arts can reduce inattentional blindness.

A principle guiding drawing, painting, or sculpting, is to draw, paint, or sculpt what we **see** contrasted with the way we think something should look. For example, before taking drawing lessons, if I were asked to draw a boat, tree, dog, or other object, I would start thinking mainly about what such an object should look like and try to draw it in that preconceived manner. Now, having benefited from drawing lessons, I draw what I **see,** that is, composition, shapes, and values, where the last item means how light or dark something is depicted.

Artists first carefully examine the object to be drawn and then, and only then, draw what they **see.** While each artist has his or her own style of converting that to pencil or brush strokes on paper or shapes in clay, careful observation drives the process. Even if the resulting artwork is not successful, the artist will still have really **seen**, probably for the first time, the object.

British Prime Minister Winston Churchill, who took up art at age forty, expressed the intensity of that "first-time" **seeing** by saying, "I found myself instinctively, as I walked, noting the tint and character of a leaf, the dreamy purple shades of mountains, the exquisite lacery of winter branches. . . And I had lived for over forty years without ever noticing any of them except in a general way" [25].

Joan Nagle draws a parallel between artists and writers by observing that both think first and then act. She recalls an incident when Roger Fry, a painter and art critic, asked a little girl how she approached drawing. Her answer: "First I have a think and then I put a line around it" [26].

Enhanced observation, or more **seeing** and, relatively speaking, less just looking, is an inevitable by-product of practicing visual arts such as drawing, painting, sculpting, and photography. **Seeing** gradually becomes habitual for artists. As Ralph Waldo Emerson noted, "The mind, once stretched by a new idea, never returns to its original dimensions," and so it is once we do visual arts.

Consider many examples of how **seeing** enhanced understanding of a system or process, solved a problem, enhanced the effectiveness of engineering work, and stimulated creativity. The following descriptions draw on my article titled "Can Creating Art Make You a More Effective Engineer" published in *PE-The Magazine for Professional Engineers* [22].

Leonardo da Vinci started pathology: As a teenager, Leonardo da Vinci spent some inspiring time in Florence, where, among other experiences, he learned about and aspired to become one of the "ingenios," the engineer/artists. Later, da Vinci, now the engineer-artist-scientist, dissected more than thirty human cadavers and many animal corpses. He conducted the first documented autopsy, in effect making da Vinci pathology's founder [27–30].

Because da Vinci **saw**, and could draw; now others could **see** what they had not **seen**. He said there are three classes of people: "those who **see,** those who **see** when they are shown, those who do not **see**," which supports the idea that doing visual art can enhance the sight of those who want to **see** more.

Philo Farnsworth invented television: In 1920, an observant and curious 14-year-old farm boy, Philo Farnsworth, **saw** neat parallel rows of crops on his uncle's Idaho farm. This caused him to think of electronically capturing an image, in a point-by-point and then line-by-line manner, transmitting it, and reassembling them into the original image. He shared the concept with his high school chemistry teacher, Justin Tolman, whom Farnsworth later credited with providing key inspiration and knowledge.

Philo Farnsworth persisted, continued his study and experimentation, and, at age 20, he demonstrated the first working television that used electronic scanning on both pickup and display devices. As you may have guessed, the person who **saw** in a farm field the possibility of electronically capturing and transmitting moving images was just beginning his creative work. He eventually received over 130 patents for his many and varied inventions [31, 32].

Alexander Fleming discovered penicillin: In 1928, Scottish biologist Alexander Fleming, while conducting research on antibacterial substances, inadvertently contaminated one of his slides with the mold penicillium notatum. Later, he **saw** a circle around the mold that was free of bacteria. Maybe the mold came through an open window or from a crumb of moldy bread. Regardless, this accident and Fleming's ability to **see** the circle led to the discovery of penicillin, as named by him. His discovery destroys bacteria that cause many types of infections and inspired scientists to develop other antibacterial drugs [33, 34].

Norman Woodland conceived the bar code: In 1948, Bernard Silver, while an electrical engineering graduate student at Drexel Institute of Technology, learned that a food store chain wanted to speed up the checkout process at their stores. He partnered with Norman Woodland, a friend and fellow graduate student, and they started to search for a system. Their first working model used fluorescent ink, but it faded and was expensive. This was the beginning of what proved to be a persistent effort with a breakthrough ending.

Eventually, mechanical engineer Woodland moved to Florida, near the beach, and continued to work on the project, now inspired by Morse code, which he had learned as a Boy Scout. He began to think about dots and dashes. One day, during the winter of 1948–49, while he was at the beach lying back in a beach chair, he stretched out a hand, put it in the sand, and pulled it back. He looked at his finger marks in the sand, **saw** lines of varying width and spacing, and conceived the bar code concept [35–37].

George de Mestral created Velcro: In 1941, George de Mestral, a Swiss electrical engineer, returned from a hunting trip with his dog and **saw** cockleburs (seeds) on his clothes and on his dog's fur. When a curious de Mestral examined the burrs under a microscope, he **saw** many stiff spines with "hooks" on the end that caught on almost anything. **Seeing** this, he thought about the possibility of repeatedly binding two materials – one with hooks and one with hoops—in a reversible manner.

De Mestral worked for almost two decades to develop a process for manufacturing his hook-and-loop fastener and bring it to market. He persisted and commercialized the now almost omnipresent fastener called Velcro. That name is a combination of two other words. They are the French words velour, meaning fabric with a soft nap, and crochet, that is, needlework in which loops of thread or yarn are interwoven with a hooked needle. The manner in which de Mestral conceived Velcro is now called biomimicry or biomimetics, that is, **seeing** something in nature and mimicking it to produce a useful product [38, 39].

Santiago Calatrava designed unique structures: The architectural and engineering firm headed by Calatrava, the famous Spanish-born engineer and architect, designs creative, now signature structures. Examples include a gently twisted Malmo, Sweden, skyscraper influenced by **seeing** the human spine; a portion of the Milwaukee, WI, art museum with bird-inspired wings that open to the sky to moderate interior illumination; and the Lisbon, Portugal, train station inspired by the **sight** of a palm tree forest.

Calatrava earned a degree in architecture and a doctorate in civil engineering and has held engineering licenses in the United States. He draws, paints, sculpts, and does ceramics. Calatrava synthesizes some of his interests with this thought: "I have tried to get close to the frontier between architecture and sculpture and to understand architecture as an art." He said this about the advantage of his being an engineer: "Being an engineer frees him to make his architecture daring" [40–42].

Wendy Crone enhanced use of microscopy: Artist Crone, PhD, is a Professor in the Department of Engineering Physics at the University of Wisconsin-Madison, where she also has appointments in the Departments of Biomedical Engineering and Materials Science and Engineering. Her research specialty is solid mechanics, which has connected her to nanotechnology and biotechnology. She does sculpting and pottery, and she also paints.

In her research, Crone uses microscopy, that is, instrumentation that provides images of objects not visible to the naked eye. She says, "The ability to **see** detail and attend to subtle changes in images is critical to my engineering research," and she believes that "these skills are enhanced by my practice of painting" (Personal communication 18 September 2018). Her view is similar to that of the German/Swiss painter Paul Klee, who said "Art does not reproduce the visible; rather, it makes visible."

As a teacher, Crone uses her art knowledge and skills to prepare visuals that enable students to understand complex concepts. Her art also supports collaboration with professional artists and scientists in preparing presentations, writing articles and papers, and designing museum exhibits [43].

***

The preceding examples illustrate the value of **seeing**. Doing visual arts can be an effective way to strengthen your **seeing** ability. Now onto the second benefit of doing visual arts, which is being a more effective composer.

5.8.3 Composing More Effectively

The word "composer," or the related word "composition," may prompt us to think of composing music. However, that's not the intended meaning here. We are in the visual arts world, where composition refers to how artists position elements or things on a painting or drawing or within a sculpture or photograph. Visual artists, intentionally or unintentionally, apply composition rules, based more on experience than theory, to engage us, set a mood, and send a message. We are not sure why, but the human brain tends to respond favorably to classical composition rules, some of which can be traced back to at least the Renaissance [44–46].

Engineers compose when creating a PowerPoint slide or other 2D visuals, deciding on headings and subheadings in a document, listing reasons to do something, selecting colors to use on the cover of a report, and deciding on the number of outcomes for a webinar.

Let's introduce and illustrate some of the composition rules. Then I will suggest ways that you, as an engineering student or practitioner, could use the rules to strengthen your communication.

- **Create a focal point:** Immediately bring the viewer to the principal object using size, color, or other visual features. Both artists and engineers often compete for attention [45–47]. Focal point examples include Mona Lisa's smile, or maybe her hands, and the white tusks or large ear of the elephant shown in Figure 5.8.
- **Rule of Thirds:** To understand the reason for this rule's name and how artists use it in drawing and painting, imagine partitioning the vertical and horizontal spaces of an artist's canvas into approximately thirds, as shown in Figure 5.9. Note the intersections—the circles.

Figure 5.8 This elephant artwork, done in colored pencil and acrylic paint, illustrates composition rules used by artists and usable by engineering students and practitioners.

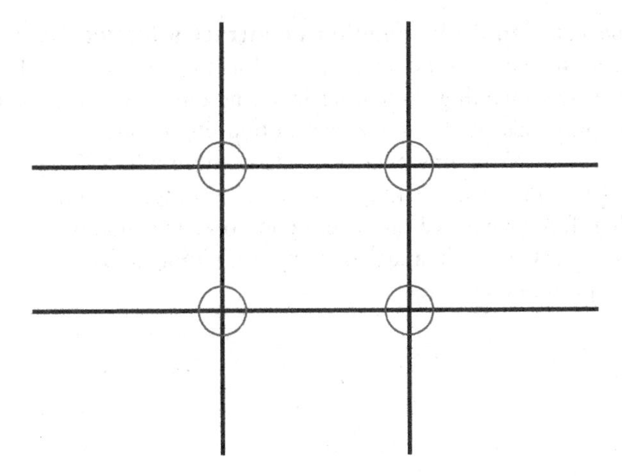

Figure 5.9 The Rule of Thirds urges artists to place the principal object off-center, near one of the circles.

Visualizing the circles, the artist places the principal object in the general location of one of the circles to assure that it is off-center [45–47]. We might think that the main object or subject should be in the middle—in the center of attention. No, the human brain usually prefers having the principal object placed at least slightly off-center, like the elephant in Figure 5.8.

- **Favor odd over even, especially three:** This rule recognizes that the human brain prefers an odd number of objects to an even number [46, 48, 49]. We can usually easily remember three words or items [47, 48]. We are often exposed to and respond to the "three" version of the odd-over-even rule, but barely notice it. Some examples:
  - 3 pieces of cutlery we use to set a table
  - 3 wishes we get from Aladdin's lamp
  - 3 inalienable rights in the U.S. Declaration of Independence
  - 3 spirits in *A Christmas Carol*
  - 3 suggestions in the safety slogan: stop, look, and listen
  - 3 primary colors—red, yellow, and blue
  - 3 colors in the U.S. flag
  - 3 musketeers in Alexandre Dumas's novel of that name
  - 3 events in the U.S. thoroughbred racing Triple Crown
  - 3 colors on a traffic light—red, yellow, green
  - 3 states of matter—solid, liquid, gas

- **Use color to evoke emotion or attract attention:** Recall the color discussion earlier in this chapter (Section 5.5.5), which shows how color has meaning, including invoking emotion. Also remember that, for purposes of this book, we define communication as "effectively conveying, from one person to one or more others, ideas, information, and feelings using asking, listening, writing, speaking, and visuals." Therefore, when we engineers compose, we should use and leverage color, just like when artists compose. Color helps us attract attention and prompt emotion.

***

Example Applications of Composition Rules

1. I created the slide very much like that in Figure 5.10 for a presentation titled "The Leader Within: The Seven Qualities of Leaders" [50]. When used in the presentation, the slide was in color, with blue text in the upper left and red text adjacent to each of the legs.

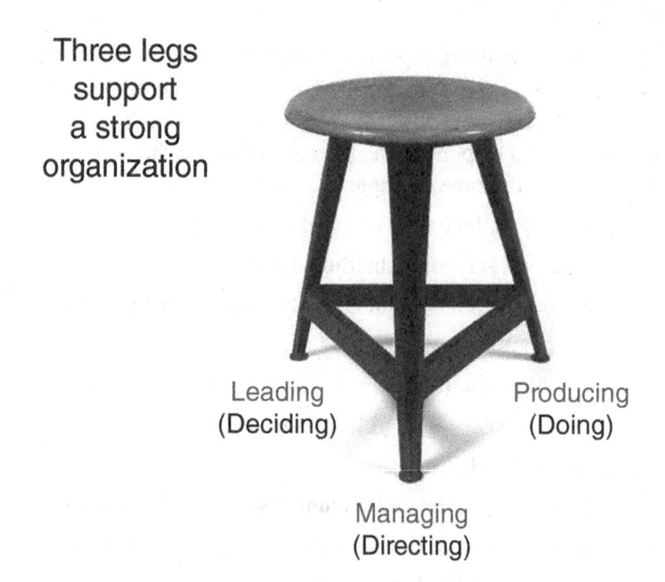

Figure 5.10 This slide illustrates all of the composition rules.

This slide, in its colored version, illustrates these composition rules:

- Create a focal point: The stool, which also suggests strength and stability
- Rule of thirds: Stool offset from center
- Favor odd or even: Three legs and "Seven" in the presentation's title
- Use color to evoke emotion or attract attention: The blue text in the original suggests authority and wisdom, and red text draws viewers to the legs

2. Figure 5.11 shows the cover of my 2021 book, in which the blue background color, in the as-published version, varies from darkest at the top to lightest at the bottom. Notice the use of the following rules of composition:

- Create a focal point: The disaster photos, especially the smoke and fire aspects, attract attention, as does the book's title with its reference to public protection.
- Rule of thirds: The two main portions of the cover—photos and title—are off-center.
- Favor odd or even: A potential reader will probably notice the three-word alliteration in the title and, once contemplated, remember it. If I could do the cover again, I would use three disaster photos.
- Use color to evoke emotion or attract attention: The blue background color suggests authority, as in suggesting that the book is credible.

3. Imagine you are an engineering graduate student writing a proposal to conduct a research project, or you are a practitioner writing a proposal to provide engineering services. In either case, you could describe your team's three strengths and the five steps you would take to produce a successful project. In doing the preceding, you are applying the odd-over-even rule. You could also design simple colored symbols for each of the three teams' strengths or for each of the five steps and use them in the proposal. Now, because of the symbols, you are also using the focal point and colors-have-meaning rules to attract attention.

5.8.4 Concluding Thoughts About Doing Visual Arts

Engineering and the Value of Doing Visual Arts

Because vision dominates our senses, enhanced visual abilities enable engineers to be more effective. By doing—not just studying—visual arts, such as drawing, painting, sculpture, or photography, we enhance our observational capabilities. We see what others don't see, which enables us to define issues, problems, and opportunities more thoroughly and more creatively resolve or pursue them. Doing visual arts also enables engineers to compose, that is, arrange ideas, principles, options, and objects for enhanced communication.

Figure 5.11 This book cover uses all four composition rules.

The engineer–art connection also works in the opposite direction. Because engineers readily "see" in three dimensions, we can represent an object in two dimensions that looks three-dimensional, and we understand the principles of perspective drawing.

Looking Beyond Engineering for Insight into the Value of Doing Visual Arts
A team of researchers tested the hypothesis that "arts and crafts avocations" correlated with the success of eminent scientists. They exhaustively studied "autobiographies, biographies, and obituary notices of Nobel Prize winners in the sciences, members of the [British] Royal Society, and the U.S. National Academy of Sciences," and they tabulated "adult arts and crafts avocations." As an example of the study's thoroughness, it included all Nobel laureates from 1901 through 2005 [21].

Researchers credited scientists for having arts and crafts avocations only if evidence indicated doing, not just studying, arts or crafts. Researchers also used data from a "1936 avocation survey of Sigma Xi members [honor society for scientists and engineers] and a 1982 survey of arts avocations among the U.S. public."

The research result most relevant to this book and consistent with and extending previous studies extending back to the early 20th century, is: Nobel laureates were about three times as likely to do arts and crafts—essentially equivalent to visual arts as used in this book—as were Sigma Xi members and the U.S. public. The arts seem to foster scientific success.

The described study focused on scientists, as indicated by the organizations used to provide data. However, given the overlap in the formal educations of scientists and engineers, some similar characteristics of individuals in the two occupations, and the fact that engineering builds on a science foundation, some of the research results are likely to apply to the engineers. I offer the study results as another way of encouraging engineers to explore the visual arts.

5.9 KEY POINTS

- A visual is something we use when conversing, writing, or speaking that appeals to sight and enables others to understand and empathize, and maybe agree with, support, and commit to our cause.
- When communicating, we should always try to use visuals because sight is our most powerful sense—it connects with more of our brain than any other four basic senses.
- We can select from two types of visuals: images and props, or what we also refer to as 2D and 3D visuals.
- We engineers and other technical professionals plan, design, construct, and operate things—products, processes, structures, facilities, and systems—that serve societal needs. Therefore, we have ready access to highly varied potential visuals, including props.
- This last chapter offers engineering students and practitioners many practical suggestions for using 2D and 3D visuals, especially when writing and speaking.
- At the midpoint of this chapter, emphasis shifts from using visuals to communicate with others to creating visuals, that is, doing art for pleasure, which also enhances our ability to see and to apply composition principles as part of engineering work.
- A comprehensive study of scientists concluded that doing art fosters the success of the most accomplished scientists. Perhaps the same is true for engineers.

The mind, once stretched by a new idea,
never returns to its original dimensions.

—Ralph Waldo Emerson, schoolmaster, minister, lecturer, and writer

---

REFERENCES

1. Merriam-Webster Dictionary. (2023). Visual. https://www.merriam-webster.com/dictionary/visual (accessed 20 May 2023).
2. Cambridge Dictionary. (2023). Visual. https://dictionary.cambridge.org/us/dictionary/english/visual?q=Visual (accessed 26 November 2023).
3. Walesh, S.G. (2017). *Introduction to Creativity and Innovation for Engineers*. Chapter 2, The brain: a primer. (pp. 29–30). Hoboken, NJ: Pearson Education.
4. Wikipedia. (2023). A picture is worth a thousand words. https://en.wikipedia.org/wiki/A_picture_is_worth_a_thousand_words (accessed 27 February 2023).
5. Yong, E. (2022). *An Immense World: How Animal Senses Reveal the Hidden World Around Us*. Introduction. (p. 11). New York: Random House.
6. Liker, J.K. (2004). *The Toyota Way: 14 Manufacturing Principles from the World's Greatest Manufacturer*. (p. 226). New York: McGraw-Hill.
7. Wilson, R. (2011). The Ohno Circle. *Indiana Professional Engineer*. January/February.
8. Berra, Y. (1998). *The Yogi Book*. New York: Workman Publishing Company.
9. Presentation Training Institute. (2020). The 6 by 6 rule for presentations explained. https://www.presentationtraininginstitute.com/the-6-by-6-rule-for-presentations-explained/ (accessed 28 September 2023).
10. Walesh, S.G. (2012). *Engineering Your Future: The Professional Practice of Engineering*. Chapter 3, Communicating to make things happen. (pp. 105–106). Hoboken, NJ: Wiley.
11. Atkinson, C. (2007). *Beyond Bullet Points: Using Microsoft PowerPoint 2007 to Create Presentations that Inform, Motivate, Inspire*. Redmond, WA: Microsoft Press.
12. Garner, J.K. and Alley, M.P. (2016). Slide structure can influence the presenter's understanding of the presentation's content. *International Journal of Engineering Education*. 32, 1(A): 39-54).
13. Wikipedia. (2023). Serif. https://en.wikipedia.org/wiki/Serif (accessed 4 June 2023).
14. Guetig, M. (2011). Harness the power of PowerPoint. *PM Network*. September. https://www.pmi.org/learning/library/powerpoint-presentations-visual-improvement-tips-2921 (accessed 4 June 2023).
15. Edwards, B. (1999). *Drawing on the Right Side of the Brain*. (pp. 230–232). New York: Penguin/Putnam.
16. Wikipedia. (2023). Color symbolism. https://en.wikipedia.org/wiki/Color_symbolism (accessed 4 June 2023).

17. Advisory Pipe Panel. (2009). *Report from the Advisory Pipe Panel to the Valparaiso City Utilities Board of Directors*. 8 January 2009.
18. Walesh, S.G. (2021). *Engineering's Public-Protection Predicament: Reform Education and Licensure for a Safer Society*. Chapter 3, Disasters: were some caused by licensure-exemption culture? (pp. 73–81). Valparaiso, IN: Hannah Press.
19. Wikipedia. (2023). Rogers Commission Report. https://en.wikipedia.org/wiki/Rogers_Commission_Report (accessed 13 June 2023).
20. Britannica. (2023). Visual arts. https://www.britannica.com/browse/Visual-Arts (accessed 26 November 2023).
21. Root-Bernstein, R. et al. (2008). Arts foster scientific success: avocations of Nobel, National Academy, Royal Society, and Sigma Xi Members. *Journal of Psychology of Science and Technology.* 1, (2).
22. Walesh, S.G. (2019). Can creating art make you a more effective engineer? *PE-The Magazine for Professional Engineers*. NSPE. March/April.
23. Chabris, C. and Simons, D. (2009). *The Invisible Gorilla: How Our Intuitions Deceive Us*. New York: Broadway Paperbacks.
24. Kaufman, S.B. and Gregoire, C. (2015). *Wired to Create: Unraveling the Mysteries of the Creative Mind*. New York: TeacherPerigee Book.
25. Churchill, W. (2013). *Painting as a Pastime*. (pp. 49-52). London: Unicorn Press.
26. Nagle, J.G. (1998). Seven habits of effective communicators. *Today's Engineer,* summer.
27. Gelb, M.J. (2004). *How to Think Like Leonardo da Vinci: Seven Steps to Genius Every Day*. New York: Bantam Dell.
28. Lankford, M. (2017). *Becoming Leonardo: An Exploded View of the Life of Leonardo da Vinci*. Brooklyn, NY: Melville House.
29. Shlain, L. (2014). *Leonardo's Brain: Understanding da Vinci's Creative Genius*. Guilford, CT: Lyons Press.
30. Wallace, R. (1966). *The World of Leonardo 1452-1519*. Alexandria, VA: Time-Life Books.
31. Brigham Young University High School. (2014). Philo Taylor Farnsworth: mathematician, inventor, father of television. http://www.byhigh.org/History/Farnsworth/PhiloT1924.html (accessed 15 June 2023).
32. Michalko, M. (2001). *Cracking Creativity: The Secrets of Creative Genius*. (p. 196). Berkeley, CA: Ten Speed Press.
33. Johnson, S. (2010). *Where Good Ideas Come From: The Natural History of Innovation*. (p. 285). New York: Riverhead Books.
34. Van Doren, C. (1991). *A History of Knowledge*. (p. 353). New York: Ballantine Book.
35. Boehler, P. (2012). N. Joseph Woodland, co-Inventor of the barcode dies. *Time,* December 14.
36. Fox, M. (2012). N. Joseph Woodland, inventor of the bar code, dies at 91. *The New York Times,* December 12.
37. Wikipedia. (2023). Norman Joseph Woodland. https://en.wikipedia.org/wiki/Norman_Joseph_Woodland (accessed 15 June 2023).
38. Bar-Cohen, Y. (Editor). 2012). *Biomimetics: Nature-Based Innovation*. Chapter 1, Introduction: nature as a source for inspiration and innovation. (p. 308). Boca Raton, FL: CRC Press.
39. Bellis, M. (2014). The invention of Velcro – George de Mestral. *About.com Inventors*. http://inventors.about.com/library/weekly/aa091297.htm (accessed August 2).

40. Encyclopedia.com. (2018). Santiago Calatrava. https://www.encyclopedia.com/people/literature-and-arts/architecture-biographies/santiago-calatrava (accessed 15 June 2023).
41. Guest, J.K., Draper, P and D. P. Billington, D.P. (2013). Santiago Calatrava's Alamillo Bridge and the idea of the structural engineer as artist. *Journal of Bridge Engineering – ASCE,* Vol. 18, Issue 10, October.
42. Santiago Calatrava – Architects & Engineers. (2018). https://calatrava.com/ (accessed 15 June 2023).
43. Crone, W.C. (2010). Bringing nano to the public: a collaboration opportunity for researchers and museums. *Journal of Nano Education.* 1-2, 102-116).
44. Calle, P. (1974). *The Pencil.* Chapter 5, Composition. Cincinnati, OH: North Light Books.
45. Hoddinott, B. (2003). *Drawing for Dummies.* Chapter 10, Planning your drawings. Hoboken, NJ: Wiley.
46. Biasotti Hooper, M.L. (2016). Composition. presented at the Venice (Florida) Art Center, January 11.
47. Krizek, D. (2012). *Drawing and Sketching Secrets.* (p. 100). New York: Reader's Digest Association.
48. Gallo, C. (2014). *Talk Like Ted: The 9 Public-Speaking Secrets of the World's Top Minds.* New York: St. Martin's Griffin.
49. Whitehouse, G. (2009). *How to Make a Boring Subject Interesting.* Pleasanton, CA: Upton and Blanding Associates.
50. Walesh, S.G. (2013). The leader within: the seven qualities of leaders. A webinar presented as part of ASCE's continuing education program, 25 September 2013.

EXERCISES

5.1 Applying the Ohno Circle: Select a process or activity that interests you and for which you are not an expert. For initial ideas, see the Ohno Circle discussion in Section 5.2. Apply the 30-minute version of the Ohno Circle mentioned there. Describe the process or activity you observed, including some photos. Indicate one to three things you learned about the process or activity and describe some potential improvements.

5.2 Alternatives to PowerPoint: Review a presentation you made using PowerPoint or other slideware. Imagine you are going to make that presentation again, but without PowerPoint. Describe and maybe illustrate the 2D and/or 3D visuals you would use.

5.3 Visual improvements you would make: Go back to the PowerPoint presentation used in Exercise 5.2 or another similar presentation that you made. Based on the suggestions for using visuals, excluding props, offered in this chapter, describe some improvements you would make.

5.4 Using props: Consider any presentation you've made that did not use props. Think of one to three ways you could have enhanced your message by using props.

5.5 Composition of the cover of your first book: Imagine that you are at a point in your engineering career where you have learned about and contributed to your chosen specialty. You are under contract with a book publisher, have completed the first draft of your book, and the publisher asks for front cover ideas.

State your specialty; choose a title, and maybe a subtitle, for your book; and compose a draft cover. Focus on cover content, composition, and color while keeping in mind your book's intended audience. Use whatever medium is convenient—this is secondary—ranging from graphic design software to colored pencils. Lessons learned from this exercise will be useful to you as a student or practitioner because of the many times you will compose, or help to compose, covers of various documents.

APPENDIX A

ABBREVIATIONS

AI	artificial intelligence
AIChE	American Institute of Chemical Engineers
AIME	American Institute of Mining, Metallurgical, and Petroleum Engineers (AIME is the correct abbreviation as used by this society)
APWA	American Public Works Association
ASCE	American Society of Civil Engineers
ASEE	American Society for Engineering Education
ASME	American Society of Mechanical Engineers
AWWA	American Water Works Association
CEO	Chief Executive Officer
CSU	Colorado State University
CV	curriculum vitae
DIP	ductile iron pipe
DNR	Department of Natural Resources
DPW	Director of Public Works
FAA	Federal Aviation Administration
FBI	Federal Bureau of Investigations
FE	Fundamentals of Engineering (as in FE examination)
F2F	face-to-face
GM	general motors
GPT	generative pre-trained transformer (as used in ChatGPT)
HR	human resources
IDNR	Indiana Department of Natural Resources
IEEE	Institute of Electrical and Electronic Engineers

The Communicative Engineer: How to Ask, Listen, Write, Speak, and Use Visuals, First Edition.
Stuart G. Walesh. Published 2024 by John Wiley & Sons, Inc.

JCC	Job Creation Committee
KSA	knowledge, skills, and attitudes
MT	Morton-Thiokol
NAE	National Academy of Engineering
NSPE	National Society of Professional Engineers
PDF	Portable Document Format
PE	Professional Engineer
PVC	polyvinyl chloride
Q&A	question and answer
R&D	research and development
RFP	request for proposal
RFQ	request for qualifications
SHW	safety, health, and welfare
STEAM	science, technology, engineering, arts, and mathematics
STEM	science, technology, engineering, mathematics
2D	two-dimensional
3D	three-dimensional

APPENDIX B

EXAMPLES OF COMMUNICATIVE ENGINEERS

B.1. INTRODUCTION

Section 1.6 presents my view that engineering students and young engineering practitioners are poised to become good to great communicators because of their:

- Intellectual gifts, persistence, and other admirable characteristics
- Ability to be inspired by and learn from exemplary engineer communicators

I prepared this appendix to introduce you to some exemplary communicative engineers. This sample consists of nine engineers, listed chronologically by birth date. I discovered them over the past several decades—you or I could find many more exemplars.

The appendix offers just enough information about each of the engineers to pique your curiosity, with the hope that you may study one or more of them. You could learn about their approach to communication and their place in the history of engineering, mostly in the United States.

B.2. JOHN ALEXANDER LOW WADDELL (1854–1938)

A globally recognized bridge-building genius and author of books and papers about bridges, railroads, lighthouses, and more. He received honorary degrees from several universities. Waddell believed that "engineers have a moral obligation to share information to improve the profession and society" [1]. Elaborating on that sharing obligation, he stated, "I have long maintained that it is utterly inexcusable for any engineer to die possessed of valuable knowledge which no one else shares" [2].

B.3. CHARLES P. STEINMETZ (1865–1923)

An electrical engineer who held over 200 patents and whose publications included 13 books and 60 articles. He advanced the development of alternating current and enabled the growth of the electric power industry [3]. Consider his thought about

The Communicative Engineer: How to Ask, Listen, Write, Speak, and Use Visuals, First Edition.

people who talk much about science—and engineering—and know little: "When it comes to scientific matters the ready talkers simply run riot. There are a lot of pseudo-scientists who with a little technical jargon to spatter through their talk are always getting in the limelight. . . The less they know the surer they are about it" [4].

B.4. HERBERT HOOVER (1874–1964)

A mining engineer and 31st U.S. President, Hoover authored 20 books with titles ranging from *American Individualism* to *Fishing for Fun* [5].

Source: The Library of Congress [6] /https://www.loc.gov/resource/cph.3a02089/last accessed September 21, 2023.

This is how he captured the creative and public-serving essence of engineering: "It is a great profession. There is the fascination of watching a figment of the imagination emerge through the aid of science to a plan on paper. Then it brings jobs and homes. . .it elevates the standards of living and adds to the comforts of life. That is the engineer's high privilege."

B.5. THEODORE VON KARMAN (1881–1963)

An expert in supersonic flight who authored or edited many books and reports [7]. Von Karman said, "Scientists define what is, engineers create what never has been" [8]. He briefly and profoundly defines the fundamental difference between scientists and engineers and reminds us why we need each other.

B.6. DAVID B. STEINMAN (1886–1960)

A civil engineer; master bridge builder around the globe; including his masterpiece, the five-mile-long Mackinac Bridge; and founder of the National Society of Professional Engineers (NSPE). He authored 600 professional papers, 20 books, and 150 poems [5].

David B. Steinman, PE, was born in New York City and raised near the Brooklyn Bridge, an experience that influenced his studies, career, and writings. Steinman's poem, "The Harp," follows and reveals how living in the shadow of the iconic bridge profoundly inspired him [9].

Five stories high above a city street
He dwelt, a child with wonder in his eyes.
For him, through winter cold and summer heat,
The sunbeams danced and stars sang lullabies.

One day as if on wings, a stranger came
And stood within the room, unheralded.
Gently he spoke, calling the boy by name:
"David, play on your harp!" he softly said.

How did the stranger guess the secret dream
That, day and night, within the child's heart burned?
Outside the window, in the sunset gleam,
Glittered the instrument for which he yearned:

A Bridge! The cables swung across the bay,
The strands that hummed like harp-strings murmuring,
They whispered to the child, "Some day. . .Some day. . ."
"There is my harp, sir. I can hear it sing!"

B.7. SAMUEL C. FLORMAN (1925-)

A construction engineer and author of seven books and several hundred articles, papers, reviews, and speeches. His second and very well-received book, *Existential Pleasures of Engineering,* describes how engineers think and feel about their work [10].

Samuel C. Florman, PE, frequently advocated a broader and deeper formal education for engineers, as illustrated here: "It seems to me that anyone who would call himself college-educated—particularly anyone who would call himself a professional—should spend some time in close communion with the great souls, the great thinkers, the great artists, of our civilization."

[11]. He also encouraged engineers to be more outgoing, partly because "engineering remains a profession characterized by anonymity" [12].

B.8. RICHARD WEINGARDT (1938–2013)

Founded an engineering firm that completed 5000 projects worldwide. He also authored hundreds of papers and presentations and a half dozen widely varying books, including *Forks in the Road: Impacting the World Around Us* [13] and *Engineering Legends: Great American Civil Engineers* [1]. Richard Weingardt, PE, encouraged engineers, by word and example, to take on leadership roles within and outside of engineering, observing, in the first-listed book, that "the world is run by those who show up."

B.9. HENRY PETROSKI (1942–2023)

A civil engineer, professor, and author of hundreds of articles and 19 books, the first of which was *To Engineer is Human: The Role of Failure in Successful Design* [14, 15]. In that book, Henry Petroski, PE, observes, "No one wants to learn by mistakes, but we cannot learn enough from successes to go beyond the state of the art." His view about the importance of communication in engineering: "Some of the most accomplished engineers of all time have paid as much attention to their words as to their numbers, to their sentences as to their equations, and to their reports as to their designs" (H. Petroski, personal communication, 6 March 2023).

B.10. MAE JEMISON (1956-)

A chemical engineering graduate who went on to become a medical doctor, astronaut, book author, and speaker—and recipient of ten honorary doctorates.

Source: NASA [16] /https://images.nasa.gov/details-S92-40463.html/last accessed September 21, 2023.

She is an effective presenter who encourages girls and young women to become scientists and engineers. Mae Jemison's advice to all of us: "Never be limited by other people's imagination; never limit others because of your limited imagination." Speaking about personal growth possibilities, she said, "Don't let anyone rob you of your imagination, your creativity, or your curiosity. It's your place in the world; it's your life. Go on and do all you can with it, make it the life you want to live" [5].

REFERENCES

1. Weingardt, R.G. (2005). *Engineering Legends: Great American Civil Engineers*. (pp. 63–66). Reston, VA: ASCE Press.
2. Waddell, J.A.L. (2002). The advancement of the engineering profession. *Civil Engineering*. ASCE. (pp. 52–55). Reston, VA. November/December.
3. Wikipedia. (2023). Charles Proteus Steinmetz. https://en.wikipedia.org/wiki/Charles_Proteus_Steinmetz (accessed 22 June 2023).
4. AZ Quotes. (2023). Top twenty quotes by Charles Proteus Steinmetz. https://www.azquotes.com/author/25739-Charles_Proteus_Steinmetz (accessed 22 June 2023).
5. Walesh, S.G. (2021). *Engineering's Public-Protection Predicament: Reform Education and Licensure for a Safer Society.* Chapter 2. Engineering excellence and engineer exemplars. (pp. 42–46, pp. 46–49, pp. 53–55). Valparaiso, IN: Hannah Press.
6. Herbert Hoover. Head-and-shoulders portrait, facing slightly right/Underwood & Underwood, Washington. Library of Congress, https://www.loc.gov/resource/cph.3a02089/ (accessed 22 June 2023).
7. Wikipedia. (2023). Theodore von Karman. https://en.wikipedia.org/wiki/Theodore_von_K%C3%A1rm%C3%A1n#Books (accessed 22 June 2023).
8. National Science Foundation (NSF). (2020). National Medal of Science: 50th Anniversary. https://www.nsf.gov/news/special_reports/medalofscience50/vonkarman.jsp (accessed 22 June 2023).
9. NSPE 60th Anniversary. (1994). *Writings of D. B. Steinman*. NSPE. Alexandria, VA, https://www.nspe.org/sites/default/files/resources/pdfs/AboutNSPE/Writings-of-DB-Steinman.pdf (accessed 8 August 2023). Poem used with permission of NSPE.
10. Florman, S.C. (1976). *The Existential Pleasures of Engineering*. New York: St. Martin's Press. A second edition was published in 1994.
11. Florman, S.C. (1987). *The Civilized Engineer*. Chapter 18, The civilized engineer: the concept. (p. 210). New York: St. Martin's Press.
12. Florman, S.C. (2007). Facing facts about the engineering profession. *The Bent of Tau Beta Pi*. Fall.
13. Weingardt, R. (1998). *Forks in the Road: Impacting the World Around Us*. Chapter 5, Those who show up. (p. 75). Denver, CO: Palamar Publishing.
14. Wikipedia. (2023). Henry Petroski. https://en.wikipedia.org/wiki/Henry_Petroski (accessed 22 June 2023).
15. Petroski, H. (1985). *To Engineer is Human: The Role of Failure in Successful Design*. (p. 62). New York: St. Martin's Press.
16. Official Portrait of STS-47 Mission Specialist Mae C. Jemison in LES. NASA Image and Video Library. https://images.nasa.gov/details-S92-40463.html (accessed 22 June 2023).

APPENDIX C

EXAMPLES OF QUESTIONS AND WHAT CAN BE LEARNED FROM ASKING THEM

C.1. INTRODUCTION

I prepared this appendix to provide you with examples of specific, pragmatic questions to ask in various communication situations commonly involving engineers. These appear as Sections C.2–C.8. My hope is that these examples will help you learn about facts and feelings in a wide variety of professional and other situations. I would not expect anyone to mechanically use all or even most of the examples. Instead, skim them, select some, and use them as the starting point for a question list tailored to your situation.

Section C.9, the last section, is unique in that it provides examples of facts and feelings learned by me and other team members because of asking questions of clients and potential clients. That section illustrates the kinds of facts and feelings you can discover, via questioning and listening, and then use to create win-win outcomes.

C.2. QUESTIONS TO ASK SOMEONE YOU JUST MET AT A CONFERENCE OR MEETING

1. So what do you hope to get out of this event?
2. I see that the keynote speaker is ________. What do you know about the speaker?
3. You mentioned that you are an electrical engineer. How did you select that specialty?
4. I'm working on ________ and having some problems, such as ________. How would you approach this, and/or do you know someone who could help me?
5. My team is completing a project, and I think other engineers would be interested in our approach. How could we get on the program for a near-future meeting of this group?
6. This is my first visit to this city. Do you know of any interesting historic sites, museums, or restaurants nearby?

The Communicative Engineer: How to Ask, Listen, Write, Speak, and Use Visuals, First Edition.
Stuart G. Walesh. Published 2024 by John Wiley & Sons, Inc.

C.3. QUESTIONS FOR A CLIENT OR POTENTIAL CLIENT ABOUT THE MOTIVATION FOR AND BACKGROUND OF A PROPOSED PROJECT

1. What/who prompted this project?
2. What is your role in the project?
3. How do you feel about the project and/or your role?
4. Is this project similar to others you have done? How? What's different?
5. What is the probability of implementation (e.g., budget/political process/permitting)?
6. What/who is the principal obstacle in implementing this project?
7. Who, outside of your organization, is or could be a strong supporter?
8. What is the main reason you are seeking outside assistance on this project?
9. Where will the project implementation funds come from?
10. What will people be saying if this project is successful?
11. Is there anything else that concerns you? [1]
12. What else may be prohibiting us from moving ahead? [1]
13. Have I asked about every detail that is important to you? [1]
14. What question should I be asking that I am not asking? [1]

C.4. QUESTIONS FOR A CLIENT OR POTENTIAL CLIENT/ CUSTOMER ABOUT HOW YOU AND THEY WOULD COMMUNICATE

1. Who makes the consultant selection decision?
2. Regular or as-needed meetings?
3. Face-to-face (F2F) and/or virtual meetings?
4. Would you prefer to conduct F2F meetings in your office, our office, or somewhere else (e.g., project site)?
5. Your and our role in front of your council/board?
6. Your and our role at public meetings?
7. Written progress reports? How often and when?
8. Liaison person(s) in your organization?
9. Major stakeholders and how to communicate with them?

C.5. QUESTIONS TO ASK YOURSELF AS YOU ARE ABOUT TO MANAGE A PROJECT [2]

1. What are three things most likely to go wrong on this project?
2. What assets/liabilities will each project team member bring to this project?
3. How might we surprise the client/owner/stakeholder and exceed their expectations?

C.6. QUESTIONS TO ASK IF YOU WANT TO LEARN MORE ABOUT AN ORGANIZATION, INCLUDING YOURS [3]

1. What made you/others "mad" today?
2. What took too long?
3. What caused complaints today?
4. What was misunderstood today?
5. What cost too much?
6. What was wasted?
7. What was too complicated?
8. What was just plain silly?
9. What job involved too many people?
10. What job involved too many actions?
11. Who consistently underperforms and needs to change?
12. Who consistently performs and warrants recognition?

C.7. QUESTIONS TO ASK YOURSELF IF YOU ARE CONTEMPLATING LEADING MAJOR CHANGE

Assume you are considering leading change in your government entity, business, university, community, professional or business society, place of worship, or other organization—all referred to below as "organization." Effecting change is very challenging. Therefore, consider asking yourself the following questions, or similar ones, before proceeding [4, 5, J. Russell, personal communication, 13 March 2006]:

1. Are you doing this primarily for the organization's benefit, or are you doing this primarily to elevate/bring attention to you?
2. What is the fundamental problem/opportunity/issue and how will you communicate it so others understand?
3. Is your commitment sufficient to deal with likely prolonged opposition and/or apathy?

4. Is the change compatible with the organization's mission and vision, or do you propose to change the organization's mission and vision?
5. Who will be positively affected by the change, and what are the "benefits" to them?
6. Who will be negatively affected by the change, and what are the "costs" to them?
7. What are the long-term implications for the organization of not changing, of proceeding in the current mode?
8. Who will not be impacted, positively or negatively, by the contemplated change but is likely to initially think that they are a stakeholder?
9. What unexpected changes could occur because of the contemplated change?
10. Is the contemplated change visionary enough to excite and engage other leaders, or are you aiming too low?
11. Can you confidently identify likely co-leaders and the reasons they will be supportive?
12. How will the core team learn more about the change process, and how will the group be expanded?
13. Who will be the principal opposition, at least initially, and why?
14. What individuals and/or organizations outside of your organization might assist?
15. Can you point to similar or related changes made elsewhere to use as examples and/or learning experiences?
16. What messages and media will comprise your contemplated communication program?
17. What are some of the major milestones and metrics needed to achieve the change?
18. What are some small successes that will demonstrate commitment and progress?
19. How will you finance and/or obtain resources for the change effort?
20. Could the contemplated change be applied on a trial or pilot basis or, once the change begins, is it irreversible?

C.8. QUESTIONS TO ASK YOURSELF IF YOU WANT TO GROW PERSONALLY AND PROFESSIONALLY

1. If you review the changes in your resume over the past five years, will you find that you had five years of experience or one year of experience five times?

2. What major personal or professional problem did you solve in the past year, and what did you learn from that experience that you could apply in other situations?
3. What was your biggest personal or professional failure in the past year, and what did you learn from that setback?
4. Who gave you major assistance, personally or professionally, in the last few years, and have you adequately thanked them?
5. What knowledge or skill have you learned in the past year, and, going forward, how will you use it?

C.9. EXAMPLES OF FACTS AND FEELINGS ABOUT NEW PROJECTS AS LEARNED FROM ASKING QUESTIONS

The following numbered facts and feelings were drawn from the experiences of teams I served on as we met with clients/potential clients to discuss possibly providing engineering services on one or more upcoming projects.

The listed results were typically not available in writing and, therefore, needed discovery during discussions. The meetings were F2F so that all participants could fully benefit from being able to ask questions, listen to answers, and observe body language (as discussed in Section 2.4.3). Each numbered item is from a separate discussion, and, of course, we learned much more—facts and feelings—during each conversation.

If, after one of the discussions, the team and I decided to submit a formal proposal to provide engineering services, that proposal would reflect what was learned—facts and feelings—during those discussions. First and foremost, the proposal would also carefully respond to a formal Request for Qualifications (RFQ), Request for Proposal (RFP), or similar document prepared and distributed by the client/potential client.

1. Want a readily accessible project manager who is technically competent and can communicate with all types of audiences.
2. Hate surprises!
3. Project team must have local knowledge and access to experts with national experience.
4. Project team must have technical expertise but no local history, presence, or interest in order to provide complete objectivity.
5. The firm, XYZ, Inc., must be on the project team.
6. Want no more public involvement than necessary.
7. Want proactive, high-profile public involvement from "Day 1."
8. Don't even mention stormwater detention/retention—we had a near drowning of a child.

9. Expect our consultant to recommend multipurpose (flood control, quality control, recreation, aesthetic) stormwater detention/retention facilities.
10. We want our project to gain national attention through presentations and papers co-authored by our consultant and us.
11. Total quality management must be a major theme in your proposal.
12. Your team better include men and women—and no tokenism!
13. During the course of this project, we want one of our engineers to learn how you operate the computer models.
14. Lots of F2F progress meetings.
15. Prefer monthly, written progress reports with minimal F2F meetings.
16. This project is really all about money—how are we going to fund whatever solutions you recommend?
17. Yes, let's consider using the more current science, technology, engineering, arts, and mathematics (STEAM) approach instead of our planned science, technology, engineering, and mathematics (STEM) approach.

The short story behind item 2: As a representative of an engineering firm, I learned I could have "one minute" with the Director of Public Works (DPW) of a community for which our firm was pursuing a large stormwater-planning project. Accordingly, as soon as I met him, I asked, "What do you like least about consultants?" The DPW's quick answer: "Hate surprises."

He went on to say he preferred F2F meetings. Although our meeting was short, it did exceed one minute. More importantly, our subsequent proposal stressed communication and included many F2F meetings and the cost of those meetings. The community selected our firm for the project. Were we chosen partly because of the question asked during the "one-minute" meeting?

Items 3 and 4 state very different expectations regarding local knowledge, which illustrates typical client preferences and idiosyncrasies, often determined only by polite and persistent questioning.

The public-involvement aspects of items 6 and 7 differ markedly. Same with the stormwater detention/retention requirements in items 8 and 9. Incidentally, the community that did not want even mentioning stormwater detention/retention was located on the largest lake in the state.

Regarding item 13, showing clients how you to do what you do naturally raises pros and cons. I have done it and never saw any harm—and it can lead to other opportunities.

Notice, in items 14 and 15, the different positions regarding meetings. This, and some of the other responses, remind us not to assume others prefer to communicate the way we do. Ask about their preferences and try to accommodate them.

Some background for item 17. A municipality and its prime consultant were designing, as part of a project, a multipurpose building that would house part of the park department's education program. They assumed the program would be STEM-based. A simple question and suggestion by a subconsultant caused them to reflect and consider the newer, more inclusive, and productive STEAM approach.

A meeting typically engages most, if not all, of the five communication modes around which this book is constructed—asking, listening, writing, speaking, and using visuals. Well-planned, executed, and followed-up meetings can lead to win-win project results. Therefore, find out early the meeting preferences of those you serve, and, if acceptable, meet when needed and meet smart.

REFERENCES

1. Fox, J. (2000). *How to Become a Rainmaker: The People Who Get and Keep Customers.* (pp. 42–45, pp. 139–141). New York: Hyperion.
2. MSA Professional Services. (2004). *Project Management Guide.* Baraboo, WI.
3. Lewis, B.J. (2000). How to manage with questions. *Journal of Management in Engineering.* ASCE. Reston, VA. January/February.
4. Maxwell, J.C. (1993). *Developing the Leader Within You.* (pp. 62–63). Nashville, TN: Nelson Business.
5. Walesh, S.G. (2012). *Engineering Your Future: The Professional Practice of Engineering.* (pp. 439–440). Chapter 15, The future and you. Hoboken, NJ: Wiley and Reston, VA: ASCE Press.

APPENDIX D

EXCERPTS FROM A PROJECT-SPECIFIC STYLE GUIDE

D.1. INTRODUCTION

Section 3.4.1 introduces style guides, noting that they help achieve consistency within a team-written document. Guides also encourage consistency among documents written within a single organization and even within a document written by one person. The consistency and clarity provided by a style guide serve writers and readers by minimizing unnecessary distractions.

Sometimes we work on a project for which a report or other major document is one of the deliverables, and there is no applicable style guide. In that case, consider preparing a project-specific guide. The remainder of this appendix presents slightly edited excerpts from a project-specific guide—it may help you and others get started on your guide.

D.2. OVERALL STRUCTURE

- Provide a preface and include acknowledgments in it
- Include an executive summary (two-page maximum)
- Include appendices
- Provide a list of abbreviations
- Etc.

D.3. WORD PROCESSOR DEFAULT SETTINGS

- Left justify
- Arial, 12 pt
- 1 inch left and right margins
- Etc.

D.4. ABBREVIATIONS

Spell out abbreviations when first used in a chapter

Do not include abbreviations in headings/subheadings

Do not spell out abbreviations that are common formal names (e.g., GRW Engineering, Inc.)

Etc.

D.5. SOFTWARE

Use Word for text

Use PowerPoint for graphics

Etc.

D.6. MISCELLANEOUS

Write in third person

Use "%," not "percent"

Use "gaging," not "gauging"

Spell out units (e.g., inches, acres, cubic feet)

Indicate quotes in text with "______________", not italics

Refer to graphics and photographs as figures (e.g., Figure 2)

Refer to lists as tables

Etc.

APPENDIX E

PUNCTUATION GUIDELINES

E.1. INTRODUCTION

Section 3.4.12 defines punctuation as the act or practice of inserting standardized marks or signs in written matter to clarify the meaning and separate structural units. In that section, I shared my favorite description of punctuation's purpose, which is herding some words together and keeping others apart.

I prepared this appendix to assist you in providing proper punctuation for your readers. It addresses this commonly used subset of forms of punctuation: apostrophe, comma, semicolon, hyphen, dash, italics, and ellipsis.

If the appendix does not meet your needs, refer to specialized books, such as *The Elements of Style* [1], *Eats Shoots & Leaves* [2], *Words That Make a Difference* [3], *The Chicago Manual of Style* [4], and *The Blue Book of Grammar and Punctuation* [5], or other books or search the internet for a solution. I used selective material from the first three books to create this appendix.

I suggest not being overly rigid in applying the guidelines because of inevitable changes and differences of opinion. For example, consider the issue of how to form a possessive with a noun that ends in "s." Should we write "Chris's approach" or "Chris' approach"? We do not have a universal agreement.

E.2. APOSTROPHE GUIDELINES

1. Form a possessive with a singular noun:

   "The bridge's north approach. . ."

   "The manufacturing plant's manager. . ." (the manager manages one plant)

2. Form a possessive with a plural noun:

   "The manufacturing plants' manager. . ." (the manager manages more than one plant)

The Communicative Engineer: How to Ask, Listen, Write, Speak, and Use Visuals, First Edition.
Stuart G. Walesh. © 2024 John Wiley & Sons, Inc. Published 2024 by John Wiley & Sons, Inc.

3. Form a possessive with an indefinite pronoun, contrasted with definite pronouns such as hers, its, theirs, yours, and ours:

   "One's rights must be. . ."

   "Somebody else's notebook. . ."

4. Omit letters:

   "It's time to submit the proposal to. . ." (it's replaces it is)

Note: *The Elements of Style* reminds us of a common error involving apostrophes, which is to "write it's for its, or vice versa." The first is a replacement for "it is," and the second is a possessive. The following sentence illustrates both: "It's a wise dog that scratches its own fleas."

E.3. COMMA GUIDELINES

1. Use to separate items in a series of three or more terms where the comma or commas replace "and" or "or":

   Instead of writing "Our options include no action or retrofitting or replacing the production line," use commas like this, "Our options include no action, retrofitting, or replacing the production line."

   Note: The last comma in a series of terms, before "and" or "or," is the serial or Oxford comma. Interestingly, British writers usually do not use that comma but United States writers do. Use whichever you wish while being consistent throughout your document. Wiley, and therefore this book, use the serial or Oxford comma.

2. Use to join two related sentences by following the comma with conjunctions such as "and," "but," "or," "while," and "yet":

   Consider these two related sentences: "The concrete bridge is more expensive than the steel bridge. It will have a longer economic life." Instead of two sentences, use a comma and conjunction to form one sentence like this: "The concrete bridge is more expensive than the steel bridge, but it will have a longer economic life."

3. Use before quoted speaking as in according to *The Northwest Times*, the mayor said, "We will have our staff design the city hall addition." Another option is to replace the comma with a colon.

4. Use comma pairs to enclose parenthetic expressions, where parenthetic means interjected explanation or qualifying explanation:

   Consider this sentence: "The best way to see the proposed solar panel site unless you are on a tight schedule is to explore it on foot." Clarify it with a pair of commas as follows: "The best way to see the proposed solar panel site, unless you are on a tight schedule, is to explore it on foot."

E.4. SEMICOLON GUIDELINES

1. Use to connect two related sentences where there are no conjunctions, such as "but" and "and," to connect them:

   Begin with these two sentences: "Henry Petroski's books are informative. They view engineering in fresh ways." Replace the first period with a semicolon to get one coherent sentence: "Henry Petroski's books are informative; they view engineering in fresh ways." Of course, we can go the other way—from one sentence to two—if a sentence with a semicolon is too long or otherwise cumbersome.

2. Use to clarify items in series:

   Try to determine the intent of the following sentence—are there four or three firms? "Proposals were received from XYZ engineers, the Chicago firm, Noitall Consultants, and Lowball Associates." Here's an unambiguous version accomplished with some semicolons: "Proposals were received from XYZ engineers, the Chicago firm; Noitall Consultants; and Lowball Associates." Three firms submitted proposals.

E.5. HYPHEN GUIDELINES

1. Use to spell numbers such as "fifty-seven."
2. Use to link nouns with nouns such as in "Darcy-Weisbach equation."
3. Use to link adjectives with adjectives such as "public-private partnerships."
4. Use to eliminate letter collisions such as writing, "de-ice," not "deice" or "bell-like shape," not "belllike shape."
5. Use to reduce ambiguity:

   For example, "A cross-section of the public" implies a good sample. In contrast, "A cross section of the public" suggests a group of angry citizens.

6. Use to indicate that a word is finished on the following line.

E.6. DASH GUIDELINE

Use to somewhat dramatically announce and frame a brief break. Some examples are quoted from this book.

> "Describe effective communication as drawing on six communication modes—asking, listening, writing, speaking, visuals, and mathematics—to convey ideas, information, and feelings" (Section 1.1.4).

"Imagine two people or several individuals—in both cases including some engineers—beginning to discuss a mutual concern" (Section 2.1.1).

"I recall looking into a box and seeing—or thinking I was seeing—the book *Learning to Write*" (Section 3.3).

E.7. ITALICS GUIDELINES

1. Use for titles of objects or major works such as albums, books, journals, movies, musical compositions, newspapers, ships, and spaceships.
2. Use to emphasize certain words, as in: "Thoroughly *define* a problem before trying to solve it."

E.8. ELLIPSIS GUIDELINE

Use an ellipsis (three dots or periods) to indicate missing words in a quote, such as the advice of sports psychologist Bob Rotella, who says, "The secret to great performances . . . is in the mind " (Section 4.5.1).

REFERENCES

1. Strunk, W. and White, E.B. (1999). *The Elements of Style-Fourth Edition*. New York: Allyn and Bacon.
2. Truss, L. (2009). *Eats Shoots & Leaves*. New York: Harper Collins.
3. Greenman, R. (2005). *Words that Make a Difference*. Section titled Giving language life. Delray Beach, FL: Levenger Press.
4. University of Chicago. (2017). *The Chicago Manual of Style*. Chicago: University of Chicago Press.
5. Straus, J., Kaufman, L., and Stern, T. (2014). *The Blue Book of Grammar and Punctuation*. Hoboken, NJ: Wiley Jossey-Bass.

APPENDIX F

EXAMPLES OF SPECIFIC FORMS OF WRITING

F.1. INTRODUCTION

I prepared this appendix to provide you with examples of specific forms of writing in support of Section 3.5. I hope that the additional detail provided by the example memorandum, letter, letter to the editor, and opinion will help you, as the need or opportunity arises, to write similar documents.

A word of caution is appropriate in today's digital world. Whatever we write and share will be in digital form, or can be digitized, and, therefore, out there in cyberspace forever. Let's not in a moment of anger, carelessness, or frivolity set ourselves up for a digital disaster years, if not decades, from now.

Walter Johnson, my high school drafting teacher, repeatedly required us to formally letter "He who thinketh by the inch and talketh by the yard should be kicketh by the foot." In retrospect, he was teaching us more than lettering. His advice about thinking before communicating applies much more in today's digital world.

F.2. MEMORANDUM

A hypothetical memorandum is used to transmit the agenda of a meeting to design team members. The three attachments mentioned are not included. The agenda is used with the permission of ASCE [1].

MEMORANDUM

Date: January 19, 2024
To: Members of the Design Team—H.O.T. Air, B. Careful, O. U. Kidd, B. Level, and U. R. Liable
From: I. M. Boss, Project Manager

Re: Agenda for Meeting 9, Wednesday, January 24, 2024, 8:00–9:00 a.m., Conference Room 2

1. Good news
2. Additional agenda items?

3. Purpose
4. Surveying (see Attachment A, for alternatives and some pros and cons) (B. Level).
   a. Discuss alternatives.
   b. Select/follow-up.
5. Design criteria (see pages from State code included as Attachment B) (U. R. Liable).
   a. Recommended course of action.
   b. Decide/follow-up.
6. Alleged design error
   a. Summary of 3/11/15 meeting (see Attachment C) (B. Careful).
   b. Discussion.
   c. Follow-up.
7. Action items
8. Next meeting

Enclosure: Attachments A, B, and C
c: Vice President I. M. Bizee (without attachments)

F.3. LETTER

Sent to the Executive Director of the Indiana Professional Licensing Agency to express concern with the possible elimination of engineer licensure in Indiana.

Stuart G. Walesh PhD PE

Consultant - Teacher - Author
3006 Towne Commons Drive
Valparaiso, IN 46385-2979
Tel: 219-464-1704
Cell: 219-242-1704
Email: stu-walesh@comcast.net
www.HelpingYouEngineerYourFuture.com

August 13, 2015

Mr. Nick Rhoad
Executive Director
Indiana Professional Licensing Agency
200 W. Washington Street, Rm. W072
Indianapolis, IN 46204

RE: JCC Recommendations on Licensing of Engineers in Indiana

Mr. Rhoad:
I am surprised and concerned to learn that the Job Creation Committee (JCC) recommended the elimination of engineer licensure in Indiana. Such action would be counter to the engineering profession's highest responsibility, mainly the protection of public safety, health, and welfare (SHW).

Allow me to share a true story as a means of illustrating the **connection between licensure and public SHW**. It's 2002 at General Motors (GM). Ray DeGiorgio, an unlicensed mechanical engineer, approved the design of an ignition switch that failed to meet GM standards. As a result, it could be accidentally shut off while a GM vehicle was being driven (e.g., driver's knee touches the ignition key), which disarmed airbags, shut down power steering and brakes, and stopped other systems.

While GM was in the process of selling about two million of the faulty vehicles worldwide in the mid-2000s, its mostly unlicensed engineers learned that some GM cars with defective switches were involved in accidents resulting in injuries and deaths to drivers and passengers. They did not recommend a recall. However, in 2006, unlicensed engineer DeGiorgio quietly led the design of a new switch that met GM specifications. Once again, he and others did not recommend a recall, so about two million defective cars remained on the road around the globe. Into the late 2000s, GM continued to receive information about the dangerous and often deadly consequences of the original switch, but the mostly unlicensed engineers still did not recommend a recall.

Finally, in 2014, after a decade of irresponsible action by GM engineers, GM recalled about two million vehicles, and various types of litigation began. As of May of this year, 107 deaths, 199 injuries, and 3350 claims are attributed to the ignition switch disaster caused mainly by the unethical acts of GM engineers. The company paid a $35,000,000 fine in 2014, faced criminal charges, and fired Ray DeGiorgio and 14 other employees including engineers. (Primary source of information: Valukas, A. R. 2014. *Report to Board of Directors of General Motors Company Regarding Ignition Switch Recalls*, May.)

In my view, the GM ignition switch disaster would not have happened, or been as severe, if the GM engineers had been licensed. Why? Because my 50 years of experience as a licensed engineer in the private, public, and academic sectors where I worked with licensed and unlicensed engineers tells me, licensed practicing engineers are distinguished from unlicensed practicing engineers in these three ways:

1. They are more likely to be technically and otherwise current because continuing education is a condition of ongoing licensure in over forty states.
2. They are required by licensing boards to be ethical, with a focus on public SHW, or risk losing their licenses.
3. They strongly view themselves as members of a profession whose paramount responsibility is the protection of public health, safety, and welfare rather than as being primarily technical employees answerable only to corporate expectations.

To reiterate, in my view, **licensure is directly connected to public safety, health, and welfare. I urge you not to eliminate engineer licensure in Indiana.**

Furthermore, I strongly suggest that Indiana consider **revoking or markedly changing its industrial exemption law** so that more engineers who practice in Indiana are licensed, which, in turn, will help to protect the public. Think of the lives that would probably have been saved and the injuries probably prevented if Michigan would not have had an industrial exemption law and if all or most of the GM engineers had been licensed. I urge rejecting the JCC recommendations, which call for lowering the bar for engineering practice in Indiana. Instead, use that discussion as a stimulant for **raising the bar** for the practice of engineering in Indiana, primarily to enhance public safety, health, and welfare.

Thank you for considering my views. I would be pleased to respond to any questions you or others may have. Please make this letter part of the formal record for the JCC's August 20, 2015 meeting.

Sincerely,
Stuart G. Walesh, PhD, PE

F.4. LETTER TO THE EDITOR

Published in the letters section of the November 2012 issue of *PE: The Magazine for Professional Engineers*.

New Organization Models

According to "Talent Search" (June, p. 20), engineering organizations are "seeing their work pick up," and therefore are "focused on recruiting and retaining the best professionals to continue to move forward." Targets include a spectrum of engineers, ranging from new graduates to seasoned engineers.

Here's a thought: Instead of going back to the status quo, think about a new organizational model, especially within engineering consulting firms. Based on my experience, many engineering consulting firms have been top heavy, relative to other professional service organizations such as law firms and dental and medical clinics. That is, engineering firms have a higher ratio of engineers to paraprofessionals than other entities. While I see no significant advantage to this traditional top-heavy consulting firm structure, disadvantages include unnecessarily high labor costs (and reduced profit) and demoralization of bright, aspiring engineers who spend too much time doing paraprofessional work.

As the economy comes back, consider retaining your best and brightest engineers, providing them with additional technical and nontechnical education and training, and having them manage and lead a growing cadre of paraprofessionals.

Stuart G. Walesh PhD, PE

Valparaiso, IN

F.5. OPINION

Published in the April 2015 issue of *PE: The Magazine for Professional Engineers*.

Add 'Brain Basics' to Your Professional Development List

Several years ago, on a whim and after an over five-decade lapse since the third grade, I returned to art by taking a drawing class—loving it—and enrolling in more classes. I initially envisioned no connection to engineering education or practice; this was simply a pleasant diversion that proved to be much more.

Stimulated by results that I never envisioned, I began to explore drawing more deeply. This led me to read Betty Edward's book, which has the dual-meaning title *Drawing on the Right Side of the Brain*. I then discovered and studied a wealth of accessible neuroscience resources. This investigative process helped me to see connections between the whole-brain approach used in visual and performing arts and improving engineering education and, ultimately, engineering practice. My research continued and included focusing on recent neurological discoveries; interacting with engineering, medical, and other colleagues; writing articles; presenting and publishing papers; conducting workshops; and signing a contract to write the book *Introduction to Creativity and Innovation for Engineers*.

We've learned so much about that three-pound marvel between our ears in the last decade or so. Examples are lateralization, the different functions of the left and right hemispheres of the brain, neuroplasticity, conscious and subconscious thinking, dominance of habits and how they form, negativity bias, gender differences, and how to care for the brain. Motivated by my "So, how can we use this knowledge to be more effective" nature, I began to explore how we engineers and others could use brain basics to work smarter, including being more productive and innovative.

This line of thinking led me to assemble a whole-brain toolbox containing various cognitive methods that can stimulate you and, more powerfully, your group, to think more deeply and widely. Toolbox contents can include the Medici Effect (how diversity drives innovation), habit replacement, mind mapping, Ohno Circle (leveraging the novice effect), fishbone diagramming, stream of consciousness writing, freehand drawing, listening to music, biomimicry, and taking time to think. All are backed by neuroscience and enable us to think outside the box. These tools and the associated whole-brain ways of thinking provide an alternative to what is known as the Einstellung effect, which is the natural tendency to consider only those approaches that have worked in similar situations. This habitual behavior means that we are often unwittingly locked into the past, and this prevents consideration of better approaches.

In spite of my good intentions, I have found very little interest in what I am learning within the engineering community and have experienced some strong pushback, including being called a charlatan. Thinking about our "thinker" and how that thinking could help us work even smarter and be even more innovative is alien to many of us. Nevertheless, I persist because the topic and its practical implications fascinate me—and might interest you and your organization. As noted by a colleague from the medical profession, the human brain is no longer the domain of academia and medicine. We engineers are smart enough to understand basic neuroscience and perceptive enough to see how we can leverage it.

I asked engineers to theorize about the cause of the pushback and researched the topic. Ideas offered or discovered include reluctance to change, being trapped in our left-brain-oriented K–college education, lack of a liberal education, fear that innovative efforts could lead to failure, discomfort with creative types, and reluctance to venture into what is viewed as a medical field.

With respect to the last listed item, you may think that we are going off on a tangent. You want to be a great engineer, not a brain surgeon. I understand that concern, but I believe that knowing a selective little about your brain will help you become a better engineer. Consider this analogy. You bought a used car and want it to get better gas mileage. Therefore, you Google gas mileage and study and experiment with selected aspects of your car, such as tire types and pressure, engine tuning, wheel alignment, and use of the accelerator. As a result, gas mileage improves. You don't have to become a certified automotive technician in order to get better mileage. Similarly, you don't have to become a brain surgeon to make better use of your brain. However, you do need to know brain basics.

I urge you to learn more about your brain. We engineers strive to stay current. Self-study, seminars, webinars, and college courses help us learn more about wide-ranging topics such as communication, analytic and design tools, ethical and legal aspects of engineering practice, and project management. Add brain basics to your professional development list. You, your organization, and those you serve will benefit.

Stuart Walesh, PhD, PE, F.NSPE, is an author, teacher, and consultant who has worked in government and business. He is co-coordinator of the NSPE Engineering Body of Knowledge subcommittee. He can be contacted at stuwalesh@comcast.net.

REFERENCE

1. Walesh, S.G. (2012). *Engineering Your Future: The Professional Practice of Engineering*. Chapter 4, Developing relationships. (pp. 135–145). Hoboken, NJ: Wiley and Reston, VA: ASCE Press. The agenda is from this chapter and used with permission from ASCE.

APPENDIX G

CASE STUDY OF SPEAKER LIABILITIES—EASY TO FIX IF SPEAKERS KNEW ABOUT THEM

G.1. INTRODUCTION

I advocate frequently recording (audio or audio/video) your presentations or having a friend/colleague attend and critique them or your practices before presenting. Little things that we do or say while speaking can significantly detract from our speaking effectiveness. We, as the speaker, are not likely to be aware of these distractions.

To illustrate this point, while attending a session at an engineering conference, I listened to and observed the moderator and the four speakers. I noted what I thought were speaking distractions as well as positive aspects of the presentations. My observations follow:

Moderator

1. Excessive use of "ah."
2. Read each speaker's background material without looking at the audience.

Speaker 1

1. Began by apologizing to the audience because some members had probably already seen/heard some of his presentation.
2. Apologized for some slides (e.g., "just word slides").
3. Looked at the screen—not the audience—90% or more of the time, even though he was using PowerPoint and he could have viewed his slides on his computer monitor while frequently looking at the audience.
4. Repeatedly said "actually," as in "this is an actual beam." These "actuals" added nothing.
5. Excessive use of "ah" and "um."
6. Spoke mostly in a monotone.
7. Exceeded allotted time.

The Communicative Engineer: How to Ask, Listen, Write, Speak, and Use Visuals, First Edition.
Stuart G. Walesh. © 2024 John Wiley & Sons, Inc. Published 2024 by John Wiley & Sons, Inc.

This was an informative presentation by a knowledgeable engineer. Excellent photographs showing applications of a newer construction technique supported it. If the speaker knew about and fixed the distractions before the presentation, his offering would have been much better. (Incidentally, marketing, a highly communicative function, was one of this engineer's responsibilities.)

Speaker 2

1. Some overly complicated slides with very small letters and numbers. The speaker apparently made these slides directly from plan sheets or pages in reports.
2. Excessive use of "um."
3. Looked at the screen too much, but nowhere near as much as Speaker 1. Again, this speaker used PowerPoint and stood right behind his computer.
4. During the question and answer period, he did not repeat questions from audience members.

Speaker 2 was also knowledgeable and discussed an interesting design and construction project supported by excellent photographs and other visuals.

Speaker 3

1. Excessive use of "ah."
2. Used serif text on some slides, which, in combination with an excessive number of words per slide, made those slides difficult to read.
3. Often struggled to find the right words.
4. Looked at the screen too much; did not look enough at the audience.
5. Exceeded allotted time.

Speaker 3 was obviously knowledgeable and presented interesting technical material. Some photos and other graphics were excellent.

Speaker 4

1. PowerPoint was not ready to go. The speaker arrived at the last moment.
2. Used serif fonts.

G.2. CLOSING THOUGHT

In preparing this appendix, I do not mean to be overly critical of the four speakers and the moderator. Like them, I have a "way to go" as a speaker.

However, in the spirit of improving our speaking effectiveness, my hope is that my observations illustrate the importance of thorough out-loud practice (Section 4.4.10) and occasionally recording our presentations and/or asking a friend or colleague to critique us. By reducing distractions, we can greatly increase our speaking effectiveness.

We cannot "fix" something if we do not know it is "broken."

Index

Note: Page numbers in "*italics*" represent figures and page numbers in "**bold**" represent tables

The Communicative Engineer: How to Ask, Listen, Write, Speak, and Use Visuals, First Edition. Stuart G. Walesh.